AF334506

The Strategic Management of Technology
in the Chemical and Petrochemical Industries

To my Wife and Sons who inspire
me to the fight...

To my Father, Mother, Brothers
and Sisters who taught me to fight...

The Strategic Management of Technology in the Chemical and Petrochemical Industries

Rogerio H. Quintella

Pinter Publishers
London and New York
Distributed in the United States and
Canada by St. Martin's Press

Pinter Publishers Ltd.
25 Floral Street, London WC2E 9DS, United Kingdom

First published in 1993

Distributed exclusively in the USA and Canada by St. Martin's Press, Inc., Room 400,
175 Fifth Avenue, New York, NY 10010, USA

Rogerio H. Quintella is hereby identified as the author of this work as provided under
Section 77 of the Copyright, Designs and Patents Act, 1988.

British Library Cataloguing in Publication Data

A CIP catalogue record for this book is available from the British Library

ISBN 1 85567 146 8

Library of Congress Cataloging-in-Publication Data

Quintella, Rogerio H.
    The strategic management of technology in the chemical and petrochemical
industries / Rogerio H. Quintella.
        p.   cm.
    Includes bibliographical references and index.
    ISBN 1-85567-146-8
    1. Petroleum chemicals industry—Management. 2. Chemical industry—Management.
3. Petroleum chemicals industry—Technological innovations—Economic aspects.
4. Chemical industry—Technological innovations—Economic aspects. I. Title.
HD9579.C32Q56   1993
660'.068'4—dc20   93-21542
                                                                             CIP

Typeset by Koinonia Ltd, Manchester
Printed and bound in Great Britain by Biddles Ltd, Guildford and King's Lynn

Contents

List of figures

List of tables

Preface

This book is about the business strategy and the technology strategy of firms in the chemical industry. As such it is certainly of the greatest interest to those who work in the chemical industry or who study it. However, this by no means exhausts its relevance. It is of interest, too, to all those who are concerned with business studies in any industry because similar problems of strategic behaviour arise in every industry.

Rogerio Quintella shows that the firms which he studied all recognized the importance of relating their technology strategy to their overall business strategy. But this is easier said than done, and especially in large organizations it presents some complex problems, which he describes in depth. He is very well qualified to interpret the case-studies which he has made since he is himself a researcher from the chemical industry with direct experience of these problems. Each of the case-studies is in itself a fascinating account of business behaviour in a specific field of technology, but the book does far more than simply present four case-studies; it is also a major contribution to the literature on business strategy, providing an excellent synthesis of the various schools of thought as well as original insights.

Economists, too, have much to learn from this highly topical piece of research. It is increasingly recognized that economists and scholars in business studies need to come together to develop satisfactory theories and models of firm behaviour. The highly abstract traditional model of profit maximization lacks the concrete realism which actual studies of firm behaviour, such as this book, convey. There is a world of difference between long-term and short-term profit maximization, and all the most difficult and interesting problems are swept under the carpet when long-term issues are neglected. Most of the problems of technology strategy are long-term and, as Quintella demonstrates, there are tensions within the firm between long-term and short-term goals which cannot be reduced to a simplistic formula. The current emphasis in economics literature on path dependence and evolutionary models finds ample justification in this excellent study.

Finally, the book relates the problems of business strategy in the chemical industry to those confronting all firms in every industry as a result of the change in 'techno-economic paradigm' associated with the diffusion of computer technology and advanced communication technology. This was not the original intention of the author, as he himself explains, but as his work went on he found that many of the problems which firms confronted were related to this change in paradigm. The decline in profitability of the bulk 'commodity' chemicals in the 1970s led to new firm strategies designed either to switch increasingly to speciality chemicals or to 'de-commoditize' the bulk chemicals by developing a much more flexible product portfolio based on new applications research around a strong technology tradition. The conclusions of the book on the ways in which chemical firms are adapting to the change in paradigm from the old 'mass-production' approach are a major contribution to our understanding of this wave of technical and institutional change pervading every branch of the economy.

For all these reasons, I commend this book more strongly and hope that it finds the wide readership it merits.

Chris Freeman
SPRU
University of Sussex

All theory, dear friend, is grey, but the
golden tree of actual life springs ever
green.

Goethe

Acknowledgements

To Howard J. Rush and Christopher Freeman for their outstanding interest, support and good will;

to Osvaldo Brasil Saldeado for his invaluable support and friendship;

to Ouida Rice for her kindness and help;

to the directors and managers of 'Cases A, B, C and D' whose good will was essential to the conclusion of this book, but whose names cannot be cited due to our agreement on confidentiality;

to CNPq – Conselho Nacional de Desenvolvimento Científico e Tecnologico (Brazil) and ACRINOR – Acrilonitrila do Nordeste S/A for the essential material support;

to Simon Garmston and PCI – Fibres & Raw Materials for their support and help.

List of abbreviations

AA	Adipic acid
ABS	Acrylonitrile butadiene styrene
ADN	Adiponitrile
AN	Acrylonitrile
BCG	Boston Consulting Group
BP	British Petroleum
BS	British standard
BS	Business strategy
CEO	Chief executive officer
FCC	Fluid catalytic cracking
HCN	Hydrogen cyanide
HMDA	Hexamethylenediamine
ISO	International standards organisation
KA	Ketone-alcohol
MFA	Multi-fibre arrangement
NBR	Nitrogen butadiene rubber
NGL	Natural-gas liquids
R & D	Research and development
R & T	Research and technology
SAN	Styrene acrylonitrile
SAPPHO	Scientific activity predictor from patterns with heuristics origins
SBR	Styrene butadiene rubber
SBU	Strategic business unit
TS	Technology strategy

Introduction

Whether like the sociologist, Marcuse, or the novelist, Simone de Beauvoir, we see technology primarily as a means of human enslavement and destruction, or whether, like Adam Smith and Marx, we see it primarily as a liberating force, we are all involved in its advance. However much we might wish to, we cannot escape its impact on our daily lives, nor the moral, social and economic dilemmas with which it confronts us. We may curse it or bless it, but we cannot ignore it. Freeman (1982: 4)

This book is concerned with the relationship between the technological and the business aspects of strategies in the chemical industry. It attempts to assess whether firms follow deliberate strategies to confront their present circumstances:

(a) In their businesses in general (Business strategies).
(b) In their product and process technologies (Technology strategies).

(i) About technology, progress and strategies

One-and-a-half centuries ago the capacity to innovate was already seen as a fundamental factor in the survival of companies, industries and nations. At that time, Marx and Engels (1848) stated that:

All old-established national industries have been destroyed or are daily being destroyed. They are dislodged by new industries, whose introduction becomes a life and death question for all civilized nations...

One century later, Schumpeter (1947) referred to this phenomenon as 'creative destruction' and employed a biological analogy, 'industrial mutation', to emphasize the continuous nature of this process. He believed that the process of destruction and dislodgement of companies and industries was a healthy social process of evolution and a powerful drive for better and cheaper products and services.

Interestingly, Hofer and Schendel (1978) used the same biological analogy to introduce the concept of strategy: 'Over the long run, only those organizations survive that serve the needs of their societies effectively and efficiently...' Within Hofer and Schendel's observation about efficiency and effectiveness lays the essence of what should be the response from any organization to the opportunities and risks created by its environment. Such observations (regardless of the observer's position within the 'spectrum' which range from cursing to blessing the innovation process) involve the most critical aspect of top management work, the planning and implementation of strategies.

The recognition of the strategic role of innovation made the chemical industry, and in particular some German companies in it, world pioneers in formal industrial research and development during the 1870s. The managerial innovation of in-house research and development made it possible to pursue strategies of innovation and a continuous search for new products and processes.

This book is a study of firm strategies in mature sectors of the chemical industry. Its background comes from the business strategy and innovation literature and is based upon empirical evidence collected through in-depth case-studies of four British chemical firms.

(ii) Innovation and the chemical industry

As a consequence of the strategies adopted, the chemical industry grew at an astonishing rate. Initially this growth was led by the development of organic synthesis and the production of dyestuffs and followed soon after by the development of polymer science and the introduction of synthetic materials, mainly in the 1930s and 1940s.

Up to the end of the 1940s technological innovation was treated by economists as an exogenous factor. The knowledge since then accumulated by the industrial economists has a major role to play in understanding the relationship between business and technology strategies. Schumpeter (1947) pioneered in this field the recognition that technological evolution has its own patterns, which were influenced by scientific and economic factors. He was also the first explicitly to differentiate invention from innovation – the first commercial transaction involving a new product, process or equipment.

The domain of catalysis and chemical engineering added the last great thrust to the chemical industry. The large-scale continuous processes led to the golden era of the 1950s and 1960s. The capital intensity of the industry made national markets too small, and each of the industry's main players spread their activities around the globe. At this time, most of chemical companies' revenue originated from commodity products.

At the end of this era, the recognition of the importance of technological advance for national economies focused the views of governments and economists on the necessity for a body of economic and managerial knowledge to boost such advances. This period also marked the beginning of business-strategy studies with Peter Drucker's *The Practice of Management* (1955). The growing complexity and scientific nature of the 'industrial environment' made it necessary to diffuse R & D activities to include almost all industries and manufacturing companies. Freeman (1982) states that: 'The growth of R&D is perhaps the most important social and economic change in twentieth century industry.' At this time not only the economic theory of innovation but also management and business studies took off as described below.

(iii) Management of innovation and the modern chemical industry

The 1970s brought a wave of serious contributions to innovation theory such as the evolutionary principles of Nelson and Winter (1977) and Abernathy and Utterback's (1979) analysis of the dynamics of product and process innovations. Among such contributions empirical studies, such as the SAPPHO project (Rothwell *et al.*, 1974), played an important role in characterizing the determinants of success and failure of innovations.

In addition to innovation theory, the literature on business strategy is an essential part of the background to this research. After Drucker's contribution described above, Ansoff (1976) created another milestone by depicting the 'dimensions and foundations' of business-strategy studies.

Olin (1972) made an important contribution to investigations concerning the management of innovation specifically in the chemical industry. In the 'market place' this period brought the oil shocks which, allied to other factors such as the maturity of technologies and markets (and endemic overcapacity), led to profit-margin erosion and tremendous changes in cost, capital and managerial structures.

During the 1980s the understanding of technological innovation as an economic phenomenon deepened, and important contributions were made by Dosi (1982) with his analysis of discontinuities in technological evolution, by Freeman (1982) through the investigation of 'science-related technologies' and by Pavitt (1984), who probed knowledge accumulation at firm level. On another side of academia, business and management schools gained further impetus through the works of Porter (1980, 1985), who created very useful models of competitive forces and value chains; Mintzberg and Waters (1985), who observed the spectrum from deliberate to emergent strategies and Ansoff (1984), who described the new managerial practice of 'strategic management'.

Not many important contributions are to be found from studies related to innovation in the chemical industry. Stobaugh (1988) and Achilladelis *et al.* (1990) are among the most important exceptions. A search for work concerned with business strategy in the chemical industry proves even more difficult, as few studies include the subject. Coombs and Richards (1991 a and b) must be highlighted as an exception, although their research was only partially concerned with the chemical industry.

In the chemical industry of the 1990s most of the major players' strategies (generally away from bulk commodities and towards speciality chemicals) apparently have 'run out of fuel'. By the end of this strategic movement, the huge size of the dominant firms, now allied to an even more complex environment, has brought a new challenge to strategic management. Although, in the past, it seemed as if almost any well-conceived research project would lead to direct results in terms of improvement to processes or products (and those improvements were directly connected with similar improvements in financial results), the situation has now changed dramatically. The decline of the process of innovation in this industry was clearly demonstrated by Baily and Chakrabarti (1988) among others. Modern innovations absorb huge amounts of investment in their research, particularly in their development, and also require considerably longer time-spans. Investment capital has become insufficient to cover all the diverse areas of interest of individual companies. At the same time, the huge size of the main players in the industry appears to be an absorber for their momentum, reducing their ability to change and adapt themselves to new and more volatile markets.

The best managerial solutions for the problems described above are still to be found. Investment in research sponsored by individual businesses or divisions is very much under pressure from their financial results, resulting in short-term initiatives with a consequential lack of potential for major innovations. At the same time, corporate research, by its very nature, is still too far from the sharp end of individual businesses, their customers, creating a paradoxical situation.

The business and innovation literatures are a natural first choice for searching for information on how to cope with the industry's present situation. It is important, however, to note that most of the researches on innovation cited above are related to the early stages and origins of innovations. There are few examples of how firms behave when these innovations become mature (Abernathy and Utterback, 1978, provide one of the exceptions). In particular, little is known about strategies of firms in this situation.

(iv) Objectives

Based upon the broad goals described above, it is possible to delineate more specific objectives as:

1 – To verify the existence of deliberate 'Technology Strategies' (TS);
2 – To analyse the interaction between technology and 'Business Strategies' (BS);
3 – To search for managerial techniques and tools for the promotion of the linkage between technology and business strategies;
4 – To search for significant differences between companies on their approach to this subject (interaction between TS and BS);
5 – To investigate whether the different approaches can be related to the type of products (commodities versus specialities);
6 – To analyse the role of corporate strategies in relation to the linkage between technology and individual business strategies;

7 – To verify whether the size and nature of corporations and 'Strategic Business Units' (SBUs) can be related to the method used to couple the two strategies (TS and BS);

8 – To search for best practices.

(v) Methodology

Owing to the considerable amount of technical and commercial data necessary for analysing business and technology strategies and to the confidential nature of most of this data, the chosen method was that of case-studies. Four case-studies were conducted using British chemical companies (the theoretical and analytical framework used in the case-studies is described in Chapter 3). The first three are large corporations, which together are responsible for more than half the British production, innovations and patents in this industry. For comparison there is also a case-study of a small firm. It is not suggested that these case-studies represent a statistically significant sample in any sense. However, they do reveal many important aspects of firm behaviour in mature research-intensive sectors.

(vi) The structure of this book

The book is divided into ten chapters as follows: Chapter 1 contains a comprehensive review of the concepts of tactics and strategies. Following an 'historical' path it begins with a discussion of Drucker's famous questions of the 1950s : 'What is our business?' and 'What should it be?' As the discussion goes deeper into the subject, the 'dimensions of the strategic problem' are defined and the three main schools of thought (managerial, sociology of organizations and theory of firm behaviour) are analysed.

Chapter 2 is concerned with the conceptual and practical 'interfaces' between the concepts of 'technology' and 'strategy' within the corporate environment. The analysis is not yet specific in terms of industry, but the examples are already related to the chemical industry. Those 'interfaces' are identified and analysed respectively as: 'technology and corporate strategy', 'technology and business strategy', 'strategy at the functional level', 'strategy for technology' and 'strategy for R & D'.

Chapter 3 describes a methodology for technology strategy analysis. This methodology can be divided into three main steps, each one focusing on a different although complementary aspect: the 'macro-level analysis' is derived from the economic theory of innovation. The concepts of 'natural trajectories' (Nelson and Winter), the 'technological paradigm' (Dosi) and the 'dominant design' (Abernathy and Clark) are the key ideas used at this level. At the 'meso-level' the main concern is the competition among different companies. Several 'models' are reviewed. The micro-level is related to the genesis of the strategies within each company and has its roots in the sociology of organizations. In order to consider the competition factor, a deep analysis is necessary and the rest of this chapter is dedicated to describing the chosen sector to be studied and the general approach of the research to this sector's individual companies.

Chapter 4 is a review of the evolution of patterns of innovation in the chemical industry from its first years to the present day. It shows the overlapping in the 1870s of the British initial leading position with German industry, which was based upon the latter's professionalization of the research activity. The growth of the American industry and its link with the oil-refining industry is explained in order to provide a better understanding of the strategic concerns of the petrochemical segment. The golden era of petrochemicals is portrayed, following an analysis of the consequences of the oil-shocks in the 1970s. This review constitutes a link between the three previous chapters and the case-studies which will follow. Chapter 4 finishes with a discussion about the possible emergence of a new 'techno-economic-paradigm' and its consequences for the chemical industry's future.

Chapters 5, 6, 7 and 8 present four case-studies on the links between business strategies and technology in the chemical and petrochemical industry. The first is related to a 'strategic business unit' within the chemical branch of one of the largest British multinationals. The product under analysis is 'acrylonitrile', an intermediate petrochemical commodity. The second is concerned with 'acrylic fibres' production by a very large materials firm. The third case-study is concerned with 'nylon and its intermediates' in another giant British multinational company. The fourth, and last case-study, is based upon a 'small' company producing synthetic lattices. The aim of this last case-study is to provide the basis for a comparison of the 'strategic management of technology' between large and small companies. All four cases were conducted through a methodology involving analysis of publications, interviews and follow-ups with top managers. The analyses are conducted by starting from each corporate strategy and directed towards the role of the specific strategic business units (SBUs). As the individual businesses are focused on, their markets and technologies are reviewed in enough detail to provide an independent view of the technical and market options of each company. The observations about each individual case-study are then compared with the models described in the first two chapters.

Chapter 9 is concerned with the comparison and analysis of the different case-studies starting with a comparison of the corporations and following with a comparison of the different SBUs. In the next step it presents a discussion about the different approaches observed in the companies studied in terms of technology strategy and its linkage with business strategies.

Chapter 10 describes the main overall conclusions from the entire research. Among the main conclusions are those concerned with the evidence of a new paradigm emerging in this industry and how it affects the relationship between business and technology strategies.

1 Strategy, tactics and planning

1.1 Introduction

The concept of strategy has been used for thousands of years as the main subject of military studies. Its use as a business concept and subject of analysis is, however, relatively recent. The first attempts at explicit analysis of the strategic problem as a business and management subject began approximately 40 years ago. This present chapter is intended to generate a broad view of the strategic problem, allowing a further discussion of its relationship with technology.

There are dozens, perhaps hundreds, of different definitions of strategy in the literature, most of which only actually address part of this complex concept. The following section looks at some of these definitions. While not necessarily the main (or most frequently used) definitions, they provide a starting point for analysis. Throughout the chapter other definitions from the generic to those that are more specific will be introduced in order to get a broad view of the topic.

1.2 The concept of strategy

Before beginning the discussion on strategy, it is important to clarify its difference from the related concept of 'policy'. In business literature it is not unusual to find the term 'strategy' used interchangeably with 'policy' (Christensen *et al.*, 1965) . On the other hand, the term is also used in business literature, as well as in the market-place, as a guideline to the solution of some types of specific repetitive problems. It is, for example, common to speak about the 'company's policy of wages or training'. Ansoff (1965, p. 119) stated that: 'Policy is a contingent decision, where strategy is a rule for making a decision'. A much more comprehensive description of 'policy' was made by Hofer and Schendel (1978, p. 23) who argued that a firm should have some decisions that are made once and then usually not changed for several years. They called such decisions 'functional policy' (for example capital structure). Other types of general guidelines for decisions that are made repeatedly over time include educational refund and inventory write-offs, which were named 'operational policies'. This view of functional policies coincides with Christensen

> Major policy in central functions, such as marketing, manufacturing, procurement, research and development, labour relations, and personnel, will be stated where they distinguish the company from others; and usually the intended size, form, and climate of the organization would be included. [op. cit., p. 95]

This last interpretation will be adopted in this work because it seems more comprehensive and consistent with the ideas to be discussed and analysed here.

1.2.1 Definitions

The *Oxford English Dictionary* (1972) defines strategy as:

> The art of a commander-in-chief; the art of projecting and directing the larger military movements and operations of a campaign. Usually distinguished from **tactics** [author's emphasis], which is the art of handling forces in battle or in the immediate presence of the enemy.

On a more management-specific view, Hofer and Schendel (1978, p. 25) defined **strategy** as: 'Fundamental pattern of present and planned resource deployments and environmental interactions that indicates how the organization will achieve its objectives.'

And Lamming and Bessant (1988) define **tactics** as: 'Modes of operation employed in pursuit of a strategic goal.'

Contrasting with the concept of tactics, definitions of strategy appear to involve very different, and sometimes contradictory, ideas.

The military definition cited above encompasses the idea of breadth, and implicitly that of timing, to differentiate strategy from tactics. Hofer and Schendel's definition represents a completely different point of view, this time related to the concept of 'pattern'.

Also from a management viewpoint, Chandler (1962) defines strategy as: '... the determination of the basic long-term goals and objectives of an enterprise, and the adoption of courses of actions and the allocation of resources necessary for carrying out these goals.' Here the view on timing is explicit, as well as matching the process with its goals and objectives, which are usually analysed as different but nevertheless closely interdependent concepts.

Chandler also sees a clear-cut difference between strategy and tactics. 'Strategic decisions are concerned with the long-term health of the enterprise. Tactical decisions deal more with the day-to-day activities necessary for efficient and smooth operations.' The timing perspective becomes clear in this statement.

Going even further into a management field of analysis of this concept, it is interesting to cite Steiner (1979, p. 4) who said that: 'To oversimplify, there are two types of management. That which is done at the top of an organizational structure is strategic management. Everything else is operational management.' This concept, although an over-simplification, provides yet another point of view, which is now linked to the idea of hierarchy.

Igor Ansoff (1965, p. 5), in his famous book *Corporate Strategy*, defines strategy indirectly as pertaining 'to the relation between the firm and its environment'. He suggests that this definition is more specific and different from a more common usage in which 'strategic' denotes 'important'.

All of the definitions cited above have different nuances given their incorporation of the ideas of 'patterns', 'timing', etc. A more popular and simpler definition was implicitly made in 1955 by Peter F. Drucker when he asked the questions: 'what is our business?' and 'what should it be?'

All of these different definitions have their limitations and advantages, though each is more appropriate for different points of view. It is not part of the scope of this book to legislate about the meaning of strategy and its different concepts and definitions. Through an analysis of their different 'dimensions', loci of application and major schools of thought, the following sections of this chapter are intended to generate a clear and broad view of the strategic problem, leading to a 'natural' concept to be adopted in this book. This concept is believed to be a 'natural' one because it is a consequence of the research objective, scope and methodology (this assertion will be discussed later on).

1.3 Dimensions of the strategic problem

As the definition of the concept differs from author to author, it follows that the study of its context and process is likely to be equally as complex. In a critical view of the strategic discussion, Tregoe and Zimmerman (1980, p. 16) state that the word 'strategy' has been used rather casually in the literature, as well as in the market-place: 'In fact it has assumed a variety of meanings, some of which confuse the **what** and **how** dimensions' [author's emphasis]. They define strategy as: 'The framework which guides those choices that determine the nature and direction of an organization.' In their work they strive to differentiate 'operational' from 'strategic', relating the first to the 'how dimension' and strategic to the 'what dimension'. These 'dimensions' seem to be closely related to Drucker's definition, stated 26 years before and clearly related to that 'what dimension'.

A more holistic vision of the strategic dimensions was described by Ansoff (1976). His analysis of the 'strategic problem' is based on three distinct dimensions: the first is described as 'processual', beginning with 'planning' and ending with 'implementation'. The second dimension was originally called the 'managerial problem' and goes from 'internal configuration to external linkages'. Last, Ansoff observed a dimension he named 'variables', starting with the 'techno-economic-informational', passing through the 'psycho-sociological' and finishing with the 'political'.

Even recognizing the limitations of the physical models when applied to the 'softness' of managerial problems, it seems fruitful to use Ansoff's 'Strategic Problem'. In Figure 1.1 Ansoff's dimensions are incorporated, but the 'process dimension' is expanded to make room for 'analysis', the 'political and psycho-social' variables are combined and the 'managerial problem dimension' is renamed 'environment'. These changes are a result of the literature review and are intended to facilitate the consideration and eventual incorporation of other authors' ideas in order to create a common language with the different schools of strategic thought discussed below.

1.4 Main schools of thought

Firms have been the subject of study for many years by a wide range of disciplines. Nevertheless, the first and perhaps main attempts have been made by economists. Many of these attempts, and certainly the early ones, are concerned with the so-called 'microeconomic theory of the firm'. However, this theory, in spite of its mathematical sophistication, contributes little to the decision-making process within the firm.

Since the early development of the 'theory of the firm', understanding of the decision-making process has been significantly improved – as has the economists' contribution to real-world management issues. What has been called the 'behaviourial theory of the firm' (within the 'industrial economics') encompasses one of the main schools of thought of 'business strategy studies'. This locus of discussion will be analysed below, beginning with the contribution made by Cyert and March (1963).

In spite of the important contribution of the industrial economists, perhaps the main school of thought related to business strategy is the 'managerial school'. Many people believe that strategy became an important subject of academic research as a result of Peter Drucker's work, mainly the classic *The Practice of Management* (1955), one of the principal contributions to this school. Side by side with Drucker's work there is the work of Ansoff, and more recently Hofer and Schendel, as well as that done by researchers at the Harvard Business School such as Andrews and Michael Porter.[1]

In analysing this school of thought, Coombs (1990, p. 5) describes the three principal loci of discussions of firm strategies:

- 'Managerial Literature', the dominant feature of which is the advocacy and belief in the efficacy of some form of long-term planning by firms;
- 'Theory of firm behaviour' where the strategies seem to be shaped by historically accumulated assets in the firm;
- 'Sociology of organizations' which sees strategic action as 'processual' and describes it as 'a continuous interaction between the context, content and the process rather than content having an objective status independent of context and process'.

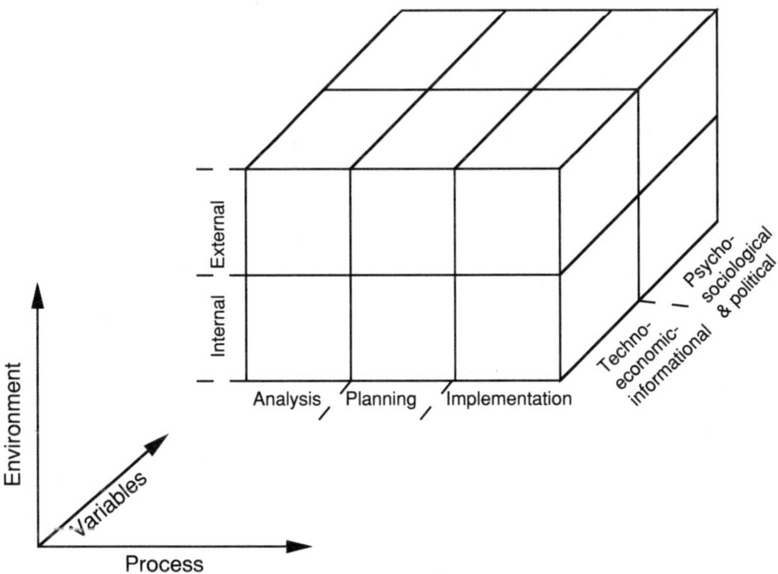

Figure 1.1   Dimensions of the 'Strategic Problem'.

The following sections are intended to provide a general view of these schools leading on to the next section, where they will be discussed in a more historical perspective.

1.4.1 Industrial economics and the 'behavioural theory of the firm'

The economic 'theory of firm' seems to assume that the objective of the firm is to maximize profits. Prices, as well as technology, are assumed to be parameters rather than variables. Some important discussion in this area can be seen in Fisher and Dornbuseh (1983) and Penrose (1959).

The classic work of Cyert and March (1963) shows some contribution also from what has been called 'organization theory'.[2] From the fusion of these two theories ('firm' and 'organization') the work of Cyert and March generated an interesting view that while considering the psycho-sociological aspects of the firms also maintained a strong economic basis. It can be illustrated by the following expression related to goals generation in the firm (Cyert and March, 1963 p. 123):

$$G(t) = a1 \times G(t-1) + a2 \times E(t-1) + a3 \times C(t-1)$$

where $G(t)$ is the company new goal, 'E' the experiences of the organization (including physical and intangible ones), 'C' an outlook of comparable organizations and the suffix $(t-1)$ refers the previous goal. It is assumed that a1+a2+a3=1, and that a3 represents the organization's sensitivity to the performance of the competitors or other similar organisations. The parameters a1 and a2 reflect the speed at which the firm revises its goals. This expression seems to be very typical and coincides with Coombs's previously cited outlook on this school.

Localizing this set of ideas and concepts into Ansoff's dimensions of the strategic problem (Figure 1.1) is a useful exercise in providing an overview of the tendencies of each school of thought. In order to do this, all these tendencies will be reviewed individually in this section and will be put together in the next section (the historical view).

The 'theory of the firm' seems to be located in the space defined by 'internal and external environment', 'analysis' and 'techno-economic-informational'.[3] In considering the 'organization theory' together with the former (see Figure 1.2), Cyert and March created a new and much more comprehensive view of the question, resulting in the so-called 'behaviourial theory of the firm'.

As explained above, it is recognized that there are limitations to physical models such as those shown in this chapter's figures. The movement and positions of each school of thought in the model should be seen as a voyage through very 'soft' and subtle frontiers. Nevertheless, the model is useful in visualizing and understanding the problem as a whole and its historical tendencies.

Perhaps the main contribution of Cyert and March's work was to open up the 'frontier' between the economic and the managerial views by admitting both economic and socio-psychological variables into the decision-making process.

1.4.2 The 'sociology of organizations' view

Henry Mintzberg is, perhaps, the main author within this school of thought. In 1978 he published a paper in which he defines strategy (p. 935) simply as: 'a pattern in a stream of decisions'. This 'fuzzy' definition enabled him to research the strategy formation in a broader descriptive context. Specifically, he was able to study two sets of strategies: those that were intended and those that were realized in spite of the original intentions. (In that work, he studied the Volkswagenwerk strategy from 1934 to 1974 in the United States, and the American government's strategy in Vietnam from 1950 to 1973.)

In a further work, Mintzberg and Waters (1985) provided a very broad framework of types of strategy-making processes. From the perfectly planned, or deliberate, strategy to the perfectly emergent one (not planned at all), they described eight stereotypes of strategies that had appeared in their empirical studies. These stereotypes provide answers to two very basic questions: 'What does it mean for an organization – a collection of people joined together to pursue some mission in common – to act deliberately?', and 'What does it mean for a strategy to emerge in an organization, not guided by intentions?'

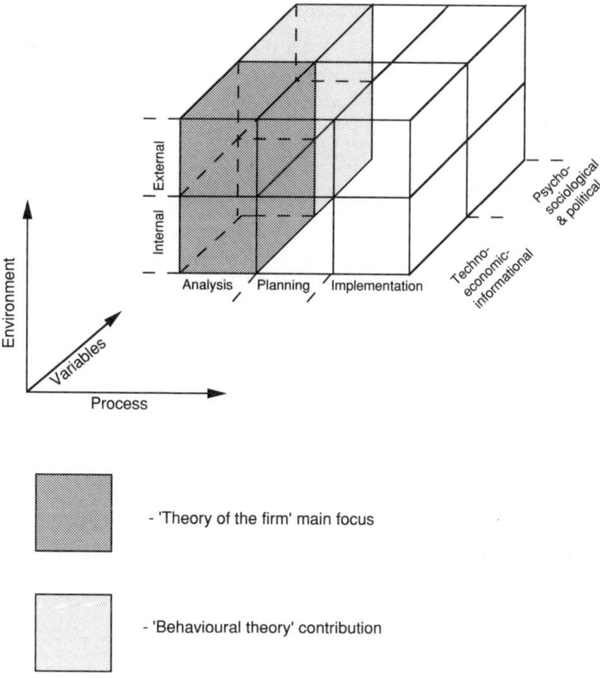

Figure 1.2 The 'theory of the firm' and the 'behavioural theory' in the strategic dimensions.

All of these different types of strategies are outlined below:

(1) Planned: they originate through formal plans, where exact intentions exist and are formulated and implemented by central leadership, using formal controls to avoid any possible surprise. Their existence depends heavily on a 'benign, controllable or predictable environment'; strategies must be deliberate.

(2) Entrepreneurial: here strategies come from a central vision again, but intentions exist as personal views. Arising from a vision of a single leader, these strategies are extremely adaptable to new opportunities and threats. Usually situated in a 'niche environment'; strategies are relatively deliberate but can also emerge.

(3) Ideological: 'strategies originated from shared beliefs', where intentions are collective and relatively immutable. Controlled normatively through 'indoctrination' and often proactive in relation to the environment; strategies are mainly deliberate.

(4) Umbrella: strategies generated from constraints. Leadership, only partially controlling organization, defines targets and goals within which other actors react based on own reasons. Rather unpredictable environment; strategies partly deliberate, partly emergent and partly deliberately emergent.

(5) Process: 'strategies originate in process: leadership controls process (hir-

ing, structure, etc.), leaving content aspects to other actors'; strategies partly deliberate, partly emergent (and, again, partly deliberately emergent).

(6) Unconnected: strategies generated in enclaves where some of the actors (loosely connected to the rest of the organization) produce their own patterns of actions consciously contradictory to central or common intentions, or in the absence of them; strategies emerge, whether or not deliberate.

(7) Consensus: strategies originate through mutual adjustment, 'actors converge on patterns that become pervasive' in the absence of central intentions. Unlike the ideological strategies, this one does not depend on 'indoctrination' but upon a more 'natural' learning from each actor to others and from the responses from the environment; strategies are, therefore, emergent.

(8) Imposed: here the main actor is the environment which dictates patterns of actions either through direct imposition or 'implicitly preemptying' organization choice; strategies are mostly emergent, 'although may be internalized by organization and made deliberate'.

Figure 1.3 shows pictorially the difference between each of these strategies. Mintzberg and Waters conclude that strategy formation (choice plus implementation) 'walks on two feet, one deliberate, the other emergent' and that management needs to direct in order to realize intentions as well as to know and to use the organisation's own pattern of action.

The contribution of this school of thought could perhaps be put in the quadrants contained by the 'planning and implementation' process and mainly 'psycho-sociological and political' variables in an 'internal environment' (Figure 1.4).

Another important author within this line of thinking can be seen in Petigrew (1989) as cited by Coombs (1990).

1.4.3 The managerial's view

In asking 'which is our business and what should it be' Drucker (1955) pointed out a critical issue whose solution is far from simple. In *The Practice of Management* he gives a broad view of management and its function. The general guidelines provided by this book are a good starting point for the huge amount of work undertaken since then. These guidelines are based mainly on a correct understanding of the market, of the customer, its values and demands. 'The question can therefore be answered only by looking at the business from the outside, from the point of view of the customer and the market.' Drucker has also suggested an unusual (for the time) view of business objectives. He considered profit maximization to be an important objective, but coming after that of the survival of the firm. While perhaps not a shocking view, compared with the then predominant view of the 'theory of the firm', it was one of the first approaches to differ from that of the industrial economists.

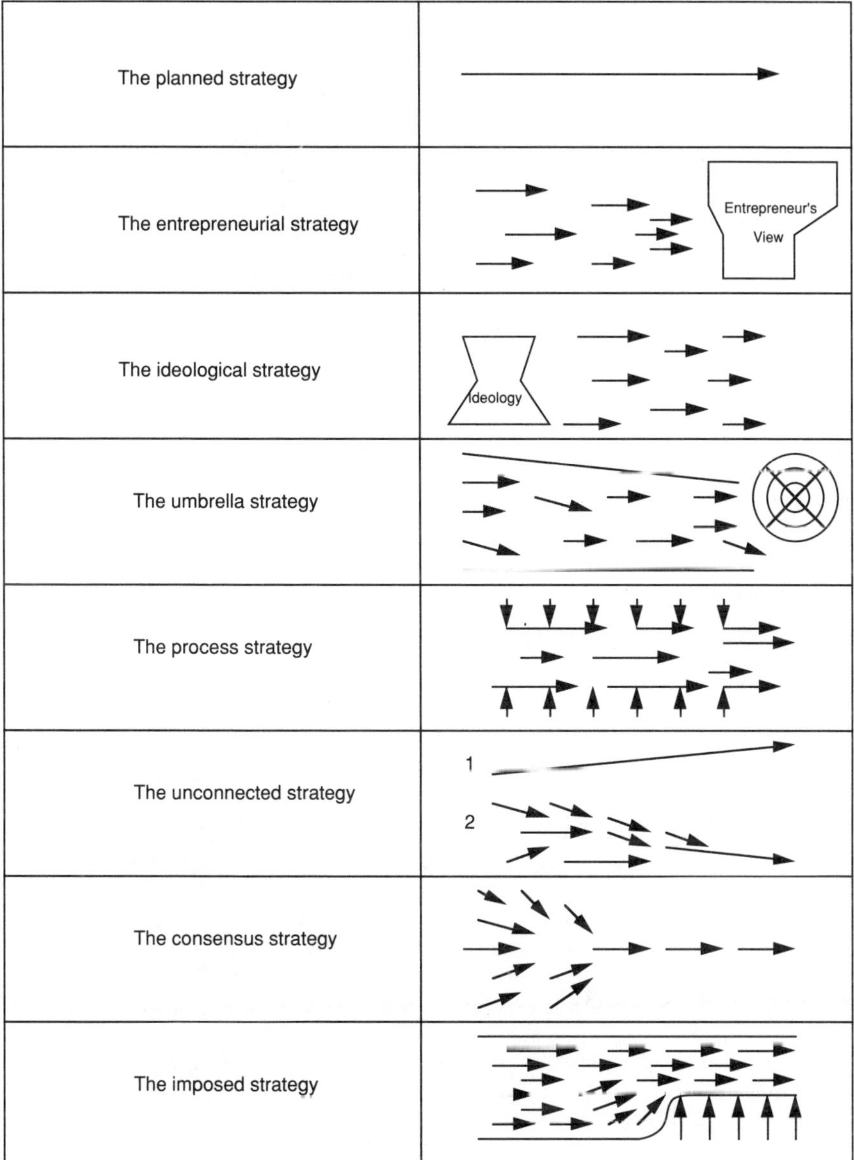

Figure 1.3   Mintzberg and Waters' taxonomy of strategies (adapted from Mintzberg and Waters, 1985).

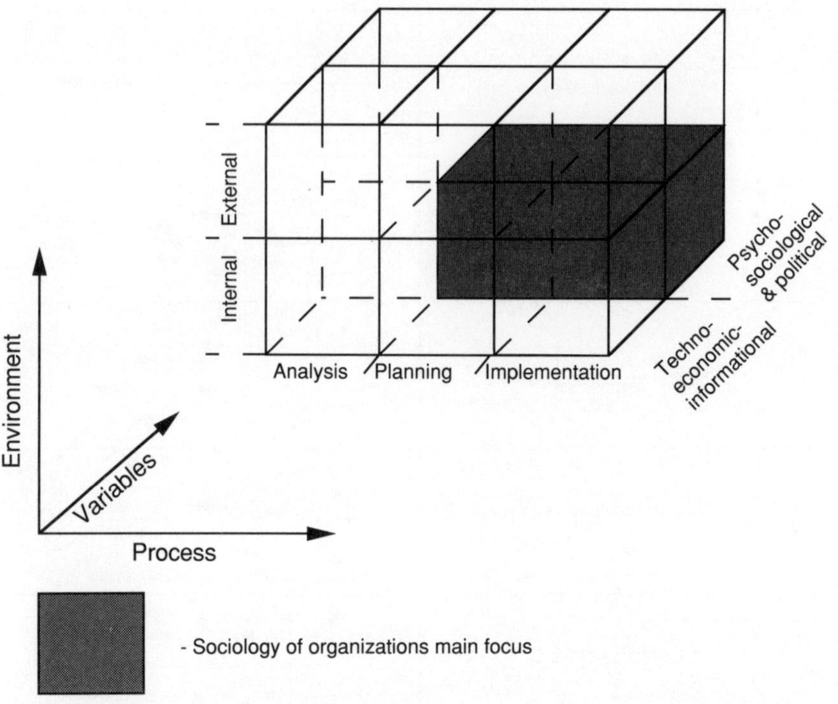

Figure 1.4 The 'sociology of organizations' in the strategic dimensions.

Later on, Chandler's (1962) study of the *History of the Industrial Enterprise* included an in-depth case-study of the relationship between strategy and structure in four giant American firms. While this work had been produced as a research work of industrial history , it became an important contribution to the 'managerial school of strategy'. His four case-studies showed a sequence of strategic steps clearly followed by structural changes in each company's organizational framework. All of these companies, in different ways, passed from centralized structures to functional ones leading to departmental organization. As described by Chandler, market, nature of resources and corporate strategy had far more effect on this history of diversification and structural adaptation than antitrust laws, taxation, labour and welfare legislation.

Another milestone in business strategic thinking was Ansoff's (1965) book *Corporate Strategy*. Here the analytical approach is different although perhaps complementary to the descriptive nature of the works previously discussed. Ansoff starts with a very comprehensive definition of the 'overall problem of the business of the firm'.[4] '... the overall problem of business of the firm is to configure and direct the resource conversion process in such a way as to optimize the attainment of the objectives.' Next he divided the 'decision space' into three parts: operating decisions, strategic decisions and administrative ones. For obvious reasons the second is the more relevant to this research. In differentiating among these decisions, he began

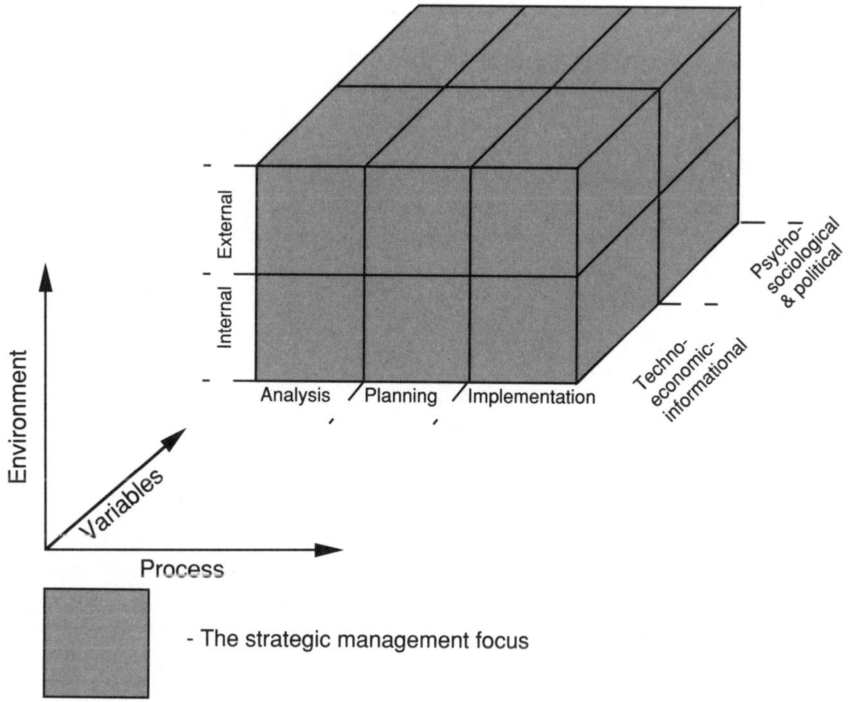

Figure 1.5 'Strategic Management' in the strategic dimensions.

by stating that they are primarily concerned with external rather than internal problems of the firm and specifically with selection of the product-mix which the firm will produce and the markets to which it will sell. The general idea is in accord with the thinking of that time and again fits well with the Drucker vision.[5]

Ansoff was one of the first to couple the idea of synergy[6] with an analysis of weakness and opportunities, and with competence profiles and competitive advantage (which eventually became the main subject of some authors within this school). Obviously, some of these concepts are present in all texts about strategy, but not necessarily the full contingent together. Ansoff also employed an analytical perspective that later became the norm.

In a later publication, Ansoff (1968) analysed the manner in which management attention is directed to needs and opportunities for strategic change, as well as its chronological steps and logistic process. This work shows the beginning of the major influence of the 'behaviourial school' on his thinking. This work concluded with a forecast of the future major steps for work on business strategy, including (p. 39) 'an extension of a Behaviourial Theory of the Firm by Cyert and March'.

It seems that the movement of Ansoff's work from its restricted position in the quadrants limited to 'techno-economic-informational' variables to one that included the 'psycho-sociological-political' dimension was not an isolated tendency but the maturation of this school of thought. This process resulted in the formation of the concept of 'Strategic Management' (see Figure 1.5) which was described in *From Strategic Planning to Strategic Management*, edited by Ansoff in 1976. (This concept will be further analyzed in the next section of the present chapter.) In his 1979 book, Ansoff was clearly concerned with the 'behaviour' of organizations which is clearly related to the 'psycho-sociological-political' quadrant of the model.

Staying within the managerial school, another important piece of work was made with an analytical approach by Hofer and Schendel (1978). This work begins with the definition of 'efficiency' and 'effectiveness'. It says that changes within a company can be of two different types: the first being those that affect the relationship between the organization and its environment while the second affect only its internal structure and operating activities. They also state that, typically, environmentally related changes affect the organisation's effectiveness, while the internally orientated changes usually have greater influence on its efficiency. Hofer and Schendel paraphrased Drucker by explaining that an organization that does the 'right things wrong' (is effective but not efficient) can outperform another organization that is doing the 'wrong thing right' (is efficient but not effective). The example used was the well-known case of the battle lost by Ford to General Motors which did the 'right things' in the 1920s by offering annual model changes and a more complete line than the very cost-efficient products of the former. From this analysis, they concluded that: 'The basic characteristics of the match an organization achieves with its environment is called strategy.'

All the analytical steps prescribed by these authors are based in their formal definition of strategy (p. 25): 'Fundamental pattern of present and planned resource deployments and environmental interactions that indicates how the organization will achieve its objectives.' From this formal definition they were able to define a hierarchy of strategies within a company as: (1) corporate, (2) business and (3) functional.[7] This strategy-formulation process is a very sophisticated and detailed one but is based on just seven comprehensive steps:

(1) 'Strategy identification' – assessment of organization's current strategy;
(2) 'Environmental analysis' – assessment of organization's competitive and general environments, identification of threats and opportunities;
(3) 'Resource analysis' – assessment of skills and resources;
(4) 'Gap analysis' – the extension of necessary changes from the current situation to the outcome which is likely from the threats and opportunities analysis;
(5) 'Strategic alternatives' – the strategic options upon which a new strategy may be built;
(6) 'Strategy evaluation' – evaluation of the strategic options in terms of values and objectives of shareholders, managers and other relevant parties;
(7) 'Strategic choice' – Selection of one or more from the strategic options.

Hofer and Schendel, once again paraphrasing Drucker, defined corporate strategy as the answer to the question 'what set of business should we compete in?' and the business strategy as the answer to 'how should we compete in the xyz business?'.

Another interesting work within this school was by Tregoe and Zimmerman (1980). Here they defined strategy as: 'the framework which guides those choices that determine the nature and direction of an organization'. While this definition is a little vague, it fits extremely well with their concept of 'driving force', which is defined as: 'the primary determiner of the scope of future products and markets'. Their research claimed the existence of nine 'strategic areas': products offered, market needs, technology, production capability, method of sale, method of distribution, natural resources, size/growth and return/profit.

Another strong conclusion was that 'one and only one of the above nine areas should be the "driving force" for the total organization'. This concept of 'driving force' seems to have some similarities with Porter's (1985) 'competitive advantage'.

Porter has been one of the most important contributors to this school. In 1965, together with Christensen, Andrews, Bower and Hamermesh he published a book, entitled *Business Policy: Text and Cases* which already incorporated some of the foundations for such later works of his as *Competitive Strategy* (1980) and *Competitive Advantage* (1985).[8] These two influential books will be covered in Sections 1.6 and 1.7.

1.5 Historical perspective

Until the end of the first half of the 19th century, there were not many companies in the world that required the services of a full-time administrator or a clearly defined administrative structure. Many of the present giant companies, however, were founded in the second half of the century, and between the 1880s and the 1920s they became large organizations. Ansoff (1976, p. viii) says that management was born and developed as an art but that during this period some academics began to apply science in management and he cites those pioneers: F. W. Taylor, Elton Mayo and Henry Fayol.[9] Most of the giant companies, mainly American, developed their present form (multi-divisional and geographically spreadout) from the 1940s. Only at that time did the 'soil become fertile' for the management school's work to go beyond academic circles. At that time long-range planning practice began to be used in big companies, eventually spreading to smaller firms. Strategic discussion, however, started only during the 1960s, becoming better known after 1970–71 when General Electric (GE) started to use it as a planning tool.

At this point, it seems worthwhile to differentiate strategic planning from its predecessor, 'long-range planning'. These two concepts have often been confused. Basically, long-range planning, as well as strategic planning, involves some forecasting of threats and opportunities. However, while the former is concerned only with present products and markets, the latter also covers new products, markets and technologies, sometimes involving radical changes like the one described by Johnson and Scholes

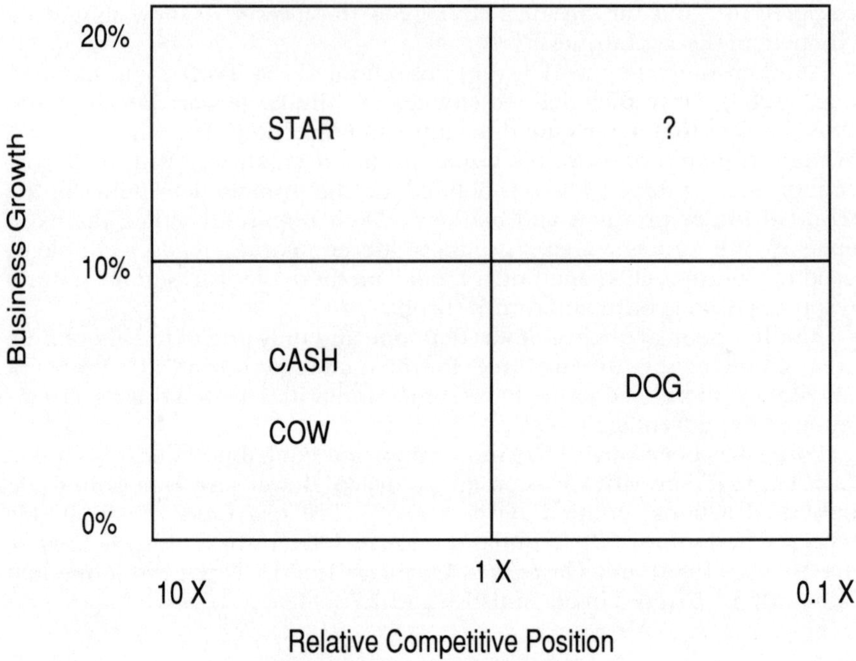

Figure 1.6 BCG's 2×2 matrix.

(1984) in the Sidlaw Industries case. Long-range planning seems more heavily based in financial concepts and, consequently, is often more bottom-up in its formulation.

From the GE 'kick-off', many companies started to use strategic planning and to build up their planning staff. At that time a variety of reasons demanded a more systematic, comprehensive and long-term approach to corporate planning. These reasons included:

- the increasing complexity of the international business and financial environment;
- intensification of global competition;
- increasing complexity of companies;
- enlargement of the scientific/technological basis of each product and process.

Wilson (1990), states that strategic planning originally occurred mainly at the business level and from 1976 at the corporate level, achieving its zenith in about 1978. This view seems to conflict with Tregoe and Zimmerman's concept of strategy, which considers strategy mainly to be a question of 'what business?' instead of 'how'. Notwithstanding this academic disagreement one of the methodologies spread throughout the 'real market place': corporate strategy analysis became popular from about 1976–77. This methodology involves the use of matrices, the most famous example

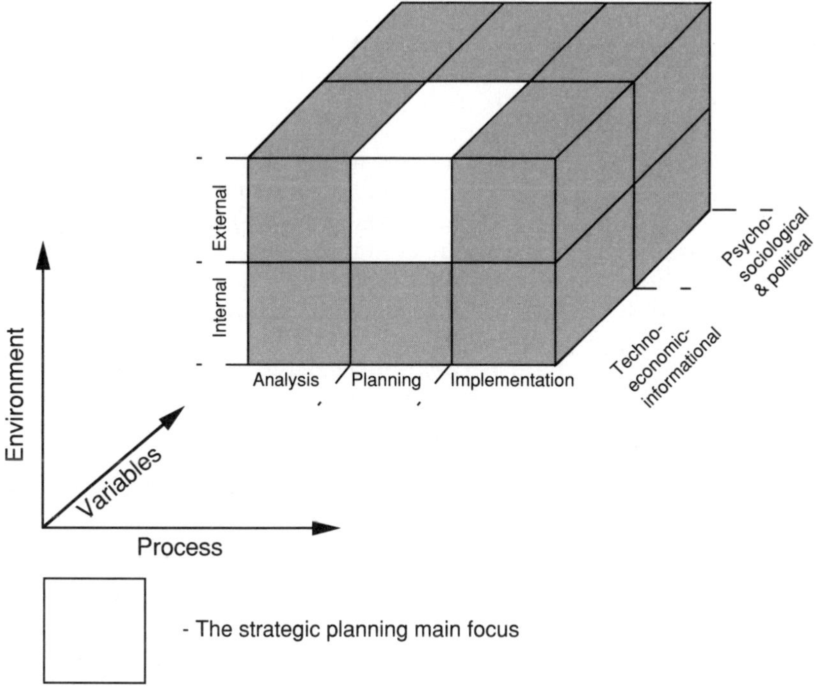

Figure 1.7   The 'Strategic Planning' in the 'Strategic Management' dimensions.

of which was constructed by the Business Consulting Group (BCG), where each of the company's businesses is plotted by its industry's rates of growth and its relative competitive position (see Figure 1.6).

Businesses plotted in the upper-left quadrant are called 'stars' because they grow quickly and are self-sustained in terms of cash flow, thus representing a good opportunity. In the lower- left quadrant are the 'cash cows' because of their low growth and high market share. This position assures low costs, low investments and consequently generates a large cash surplus. The businesses within the lower-right quadrant are called 'dogs' because they are not very profitable, having a relatively high cost and low or nil cash flow. These businesses should be divested.

The last classification is in the 'question marks' or 'wild cats' quadrant companies that have the worst cash flow owing to low market share. These businesses should be grown into a 'star' or divested.

At the same time many other matrices were developed, some of the more famous being made by other consultancy companies such as Booz Allen, Arthur de Little and McKinsey (as well as Shell and General Electric). The quantity and notoriety of these consultancy's methodologies show the importance of this process. At that time the 'consultancy boutiques' became a 'meeting point of executives', Ansoff (1976) nevertheless pointed out that: 'The technology of strategic planning has now been around for some 15 years ... Today only a handful of leading firms employ genuine strategic planning...' Possibly because of this gap between theory and practice, the period between 1979 and 1983 saw a counter-revolution. According to Wilson (1990), 'during this period, disenchantment and cost-reduction measures combined to decimate many corporate planning staffs'.

In the academic world at that time, a clear movement could be observed towards a more holistic view. This movement was even clearer in the 'managerial school' and led to the concept of 'Strategic Management'. The broadness of this concept is shown in Figure 1.7, where the relatively small scope of the strategic planning contrasts with the complexity of the 'strategic problem' as a whole. The 'strategic-management' concept involves not only 'techno-economic-informational' variables but is also heavily concerned with 'psycho-sociological and political' ones. The 'sociology of organizations' now seems to have become amalgamated with a focus on the managerial, and the interaction between 'content, context and process' (as mentioned by Coombs, 1990) is not only recognized but actually used during the 'analysis, formulation and implementation' process.

Wilson goes on to describe a series of reasons for calling that period one of 'disenchantment' but nevertheless concludes that the utilization of strategic thinking is once again growing. He cites Porter (1987):

> The need for strategic thinking has never been greater ... There are no substitutes for strategic thinking. Improving quality is meaningless without knowing what kind of quality is relevant in competitive terms. Nurturing corporate culture is useless unless the culture is aligned with a company's approach to competing. Entrepreneurship unguided by strategic perspective is much more likely to fail than succeed. And, contrary to popular opinion, even Japanese companies plan.

1.6 Corporate strategy

While there are several different definitions and interpretations of the concept of strategy, and particularly business strategy, there seems to be greater consensus about corporate strategies. In general, it is accepted that something which makes a corporation more than the sum of its individual businesses can in some way be linked to corporate strategy.

This difference between the results of a corporation and those of single independent units is defined as synergy and was largely employed to justify intensive diversifications that occurred mainly from the 1960s until the early 1970s. As observed by Porter (1985): 'synergy was a nice idea but rarely occurred in practice. Instead of synergy, the answer seemed to lie just in decentralization.' Still, according to Porter, the combination of disenchantment with synergy and the focus on decentralization has reinforced the view that 'portfolio management' is the essential task of corporate strategy. However, the importance of the concept of synergy lies with benefits generation, that surpluses the inherent costs of lost independence from each unit.[10] An interesting example of that 'misuse' of the concept of synergy comes from the chemical and petrochemical industry, which diversified particularly quickly during the 1980s. That movement (mainly towards speciality chemicals) has been frequently associated with some type of technological synergy by some of their executives. Most of the time this synergy was justified through the common aspects of the 'chemical technology' in both industries (petrochemical and the specialities). The real reasons for the movement away from bulk chemicals, however, are in fact associated with several other factors (discussed in Chapter 4). The importance of technology in companies' strategies, as well as the importance of strategies in technological change itself, are discussed in the next chapter.

1.7 Strategy at the business level

As shown by Chandler (1962), the typical multi-divisional company was born during the first decades of this century as an evolution from a single business (often passing through a phase of being a conglomerate of independent businesses). With the maturing of strategic studies, the necessity for a better visualization of the different businesses within a corporation became stronger. As a result the concept of 'Strategic Business Unit' (SBU) began to be widely used during the second half of the 1970s.

The *Macmillan Dictionary of Business and Management* (1988) defines SBU as: 'The management concept of dividing up a company into finite business units, each responsible for progress and profitability.'

Despite this simple definition, it is not always a simple task to identify the different units in a highly integrated industry (the chemical industry for example). In order to do that, it is important to have an internal and an external view of each business. For the former, it is necessary to identify – in each unit – significantly different internal activities and an absence of common sources and cost structures. For the latter, it is necessary to know the products and the characteristics for which each group of customers is looking. Such conditions should allow the different SBUs to develop their own strategies. Porter (1985) states: 'The strategic mandate to business units should be to achieve competitive advantage.'

Before analysing this work, it will be necessary to review some of Porter's basic concepts.

(1) Industry structure and the competitive forces: Figure 1.8 shows his five competitive forces (bargaining power of suppliers, bargaining power of buyers, threat of new entrants, threat of substitutes and rivalry within industry).

(2) Generic competitive strategies: according to Porter, there are four generic strategies determined by a firm's competitive advantage and competitive scope. Companies whose competitive advantage leads to 'lower cost' will (usually) have a 'cost leadership' generic strategy when their competitive scope is 'broad'. If their scope is 'narrow', the strategy will be termed 'cost focus'. But if their competitive advantage leads to differentiation, there will be another two possibilities: 'differentiation' when combined with 'broad scope' and 'differentiation focus' when the scope is 'narrow'.

Besides these two concepts, widely utilized in the literature as well as in the market place, Porter also defined the concept of 'value chain'.

(3) Value chain: the 'value chain' may be used as a tool designed to make a systematic examination of all activities a firm performs (and their interaction) in a way that allows the analysis of the sources of competitive advantage. The value chain desegregates a firm into its strategically relevant activities in order to understand the behaviour of costs and potential differentiation. 'The appropriate degree of desegregation depends on the economics of the activities and the purposes for which the value chain is being analysed.'

From these concepts, an in-depth analysis of the business value chain (through an iterative method) allows the planner to identify its weaknesses and capabilities. This iterative process is expected to generate clear ideas of how to obtain competitive advantage and how to sustain it, avoiding possible pitfalls.

These concepts, ideas and techniques require a deeper analysis, which will be made in the chapter on 'analytical framework'. Complementing the discussion presented in this chapter, some considerations will be given to strategy at the functional level in Chapter 2.

1.8 Conclusion

All the concepts and definitions of strategy discussed and analysed in this chapter are just a small part of a complex ocean of concepts, definitions, ideas and approaches. The terminology is far from precise and has different meanings for different authors and schools of thought. The discussion presented above was intended to provide a broad view of the main currents of thinking. From these fuzzy ideas this concluding section is intended to define the concepts to be adopted in the rest of this book.

In spite of the large number of definitions of strategy, it is clear that all of them are based on just a few central concepts such as:

- 'Importance'; explicitly used by Pearson (1990), implicitly by the *Oxford Dictionary*.
- 'Time'; used explicitly by Chandler and implicitly by Hofer and Drucker.
- 'Pattern'; used explicitly by Hofer and Mintzberg, implicitly in Tregoe and Zimmerman's work.
- 'Environmental interactions'; explicitly used by Ansoff and later by Hofer.

In some way, almost all strategy definitions can be related to one or more of these concepts.

As Pearson wrote:

Strategy is not about planning, but about thinking and doing. It is not a technique, but a way of managing the business according to a strategic understanding and perspective.

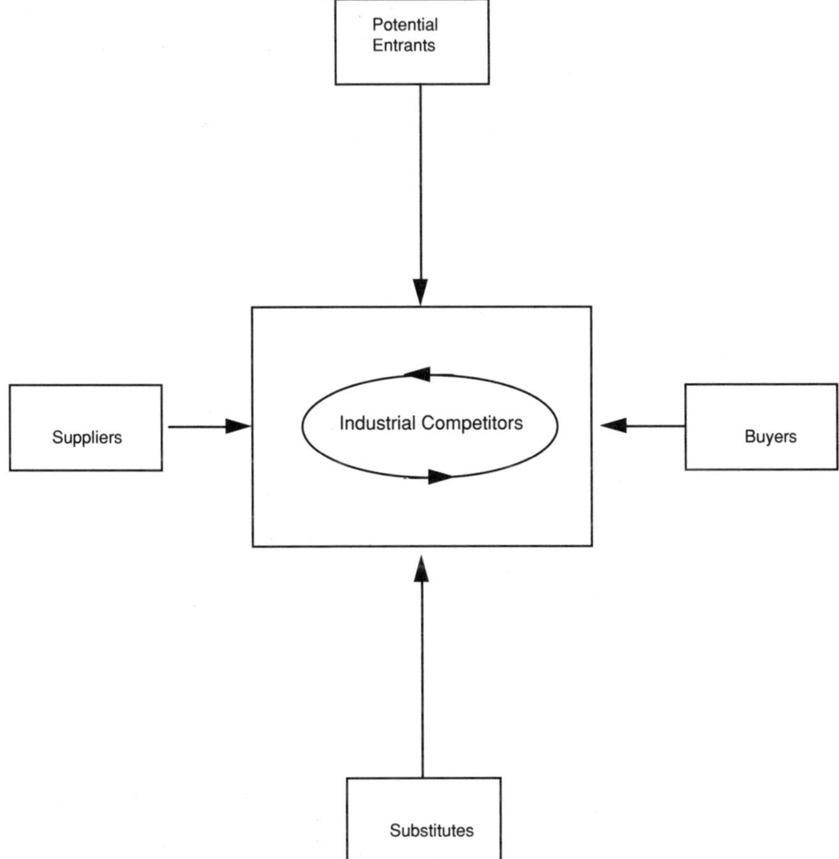

Figure 1.8 Porter's 'Competitive Forces' chart (adapted from Porter, 1985).

There is no single best strategic perspective – strategic thinking can be structured in various ways. None provides the answers to every situation, but in most approaches there is something of value. They have relevance in particular circumstances or to do a particular job. If they are understood and the circumstances are recognized, it may be possible to formulate a strategy which is effective and incisive. [1990, p. 20]

The historical perspective of strategic thinking discussed above shows clearly that this broad view was often disregarded, but it is also evident that strategy is a process and should be a continuous process of thinking and planning about survival. All the techniques should be seen merely as tools to analyse and understand the whole process.

It is commonly agreed that strategy is intrinsically linked to importance, but the problem is that even importance is relative. All things can be of extreme importance or of no importance, depending basically on the perspective of the observer. Considering that the main objective of this work is to improve the understanding of how business strategy and technological activities influence each other in a company, it is important to clarify the perspective and concepts to be used in this analysis.

The discussion presented in the previous sections of the present chapter showed a distinct movement, mainly within the 'managerial school', from a view based only on 'techno-economic-informational' variables to a more holistic view of the 'strategic problem'. This tendency towards a holistic concept of strategy seems actually to be a new step for all the schools of thought. It seems to be necessary and natural, therefore, to use here some views that do not ignore this drift. It is also believed that the actions employed to accomplish a goal may strongly influence the time span in which its objectives are achieved. Accepting this assumption, definitions used in this book will include an environmental dimension instead of temporal ones. In order to utilize, at least partially, the wisdom charted by the three schools of thought, these definitions should not contain a hierarchical concept (that goes against the sociology of organizations' view as well as that of the 'modern managerial school'). At the same time, it should be considered necessary to understand the strategy of several different companies at the same time. For all these reasons, it seems natural to use the following definition for strategy:

> A pattern of resources and competencies deployed to achieve a goal or objective, which involves environmental interactions.

This definition clearly distinguishes between the goal or objective and the means to be employed for its achievement. In accordance with this definition, strategy is also understood to be independent of any planning or timing.[11]

The concept of 'pattern' allows a researcher to capture each company strategy not only from the view of its planners and managers but also from external observers – customers, consultants and trade associations. It also includes the 'emergent' component of companies' 'planned strategies'.

The same considerations could be also applied to the following definition of tactic:

> A pattern of resources and competencies deployed to achieve a goal or objective, which does not involve environmental interactions.

To clarify the utility of such definitions, let's give some considerations to their uses. The four stereotypes of generic strategies defined by Porter – cost-leadership, differentiation, cost-leadership-focus and differentiation-focus – seem to fit well with the definition provided above. To become effective, all of them require a pattern of resources and competencies necessarily involving interactions with the business environment (in this case, the market). At the same time, these stereotypes seem to be easily identifiable through contact with their environment and even directly with each company as they do not involve any type of secrecy.

Notes

1. An important contribution to the strategic-planning process was also made by consultancy companies such as Boston Consulting Group, Booz Allen, Arthur de Little and others.
2. These were divided into three branches: first, the 'sociological', whose 'fathers' were Weber (1962), Durkheim (1933) and Pareto; second, the 'social psychological' with its emphasis on an 'efficiency' criteria and last the 'administrative', which focuses on the relationship between the executive and the organization.
3. It seems that only the economic part of this variable was considered by the authors previously cited. The consideration of the 'technical' variable only became an object of study in the last few decades within the group of economists concerned with 'industrial innovation' as discussed later.
4. Actually this definition seems comprehensive enough to describe the 'managerial problem', even for non-profit-orientated organizations.
5. The strategy components were defined by Ansoff as: (a) Product-market scope; (b) Growth vector; (c) Competitive advantage; (d) Synergy.
6. In business studies, it means joint effects resulting from addition of new product-markets, techniques or any other activities within the firm where the result is larger than the summation of the parts.
7. Here, they recognize the possible existence of another level of strategy (termed by Ansoff, 1977 'enterprise strategy'). Hofer and Schendel forecast that this concept would emerge in the following two decades to address the question, 'how can we maintain the legitimacy of the organization?' This forecast was successful, as this type of preoccupation became very evident in several companies. For example, Cookson and Jackson, (1991) in the *Financial Times* said: 'At the same time the company will be preparing to fight a propaganda battle for the hearts and minds of the British people. According to opinion polls many of them see the chemicals industry, epitomized by ICI, as a polluting monster.'
   This preoccupation is very clear also in several giant chemical companies which attempt (beginning with their annual reports) to associate their name, not with pollution, but with a commitment to environmental preservation. For example see the annual reports of: Dow (1989), ICI (1987–90), BP (1988–90).
8. A third important publication is Porter's *The Competitiveness of Nations,* although this is outside the scope of this present research.
9. Christensen *et al.* (1965) also wrote about the pioneer work done by A.W. Shaw, in 1911 the first professor of business policy at the Harvard Business School.
10. From a different perspective, the necessity for synergy can be seen as the necessity to create diversification more cheaply than can be done by one of the shareholders acting alone.
11. It is necessary to explain, however, that it is believed that planning can significantly improve the results and the effectiveness of a strategy. The fact that this component is disregarded in the definition does not mean that it should be equally disregarded in its use. On the contrary, it explains why this work is much more closely linked to the 'managerial school' than to any of the others.

2 The interface between strategy and technology

2.1 Introduction

The preceding chapter explored the idea of strategy at the business-unit and the corporate levels. This chapter is intended to identify the possible relationship between the concepts of 'strategy' and 'technology' at each of the different levels. Following this theoretical analysis at the interface, Chapter 3 will address the practical meaning of each of these interfaces and evaluate possible techniques or tools that can be used to synchronize technology-related activities with the broader strategies of an industrial enterprise. While both chapters are directly related to the general scope of technology strategy, the present one is more introductory in its nature and most of the authors have links with management literature. Chapter 4 is intended as a review of tools and techniques, and several of the authors there cited are associated with the literature of industrial innovation.

The importance of the linkage between technology and business strategy was well exemplified in recent work by Szakonyi (1990), where he states that the absence of R & D leverage was responsible for the lack of competitiveness of American companies in comparison to Japanese. He observed that the main reason for all these problems was poor coordination between R & D and business planning.

In analysing the relationship between the concepts of strategy and technology, there are three different loci of analysis: the first is the analytical interface between technology and the overall corporate strategy. The second refers to the interface at the business level and the third is related to a technology strategy.[1]

2.2 Technology and corporate strategy

Corporate strategy is managed mainly through the business portfolio and has two main issues: the first is the exploration of synergy among the different businesses (which should make the corporation more attractive to its shareholders than a group of individual companies). The second is the management of the portfolio in such a way that it can optimize the probability of companies' long-term survival.

Main sources of synergy in a corporation are its technology and expertise in general. In analysing the importance of technology upon the corporate strategy, Fusfeld (1978, p. 54) introduces the concept of 'technology elasticity'. He compares technology elasticity with the concept of 'price elasticity' used by economists and illustrates this with the case of Black and Decker. This company was originally orientated to the market of commercial and industrial construction tools, having an insignificant penetration in the home market. The recognition of different technical demands and price constraints in the two markets led the company to differentiate their production. According to Fusfeld, such a strategic move was possible because of their technology elasticity, which allowed the application of the 'same technology' to different segments of the market.[2] In his view, elasticity is missing from most research and development plans and could effectively help in a strategic use of technology in corporate planning.

This concept of technology elasticity exemplifies what can be achieved through the exploration of technology as a source of synergy. A further important aspect to be analysed at the corporate level is how a company approaches a 'technology strategy'. It is generally accepted that there are two types of approach: the 'traditional', where companies look for technologies that can help to maintain or 'attack' a specific market; and the 'competence-based', where the analysis is inverted whereby from its existing competencies a company looks for new applications and markets. (These two types of approach can be identified at the business level as well.) An exploration of the so-called technology elasticity is clearly within the competence-based approach.

Technology can certainly be a very important source of synergy mainly because of its intangible nature. Another good view of such synergetic potential can be seen in Cookson's (1991) report about ICI's review of its R & D activities. In an interview with the director of research and technology, the article explored the possible advantages of separating ICI's pharmaceutical division from the rest of the corporation and merging it with another pharmaceutical company. The main argument used by ICI's director to deny that such advantages existed was the synergy that the company could get from its different businesses:

> 'Fungicides are the best example of two-way exchange of ideas in ICI', he says. 'Pyrimidine compounds, used as fungicides for cereals, come originally from pharmaceutical research. Conversely triazole compounds, developed by the agrochemical business, have given pharmaceuticals a potent systemic fungicide' ... If ICI is to get the benefit of its investment in science, that has to transcend business boundaries.

This example seems to indicate that the exploration of technology synergy is in someway related to the 'competence-based approach'. However, such a conclusion is not clear-cut, as can be justified by the discussion below.

The second important issue in the report about that chemical corporation is the observation that such a giant company does not have a central corporate laboratory for conducting basic or strategic technology development (the adoption of this type of approach seems to be a common feature of the R & D strategy of several companies as shown by Rosenbloom and Kantrow (1982) who provide examples from Xerox, Du Pont, Monsanto, RCA and Exxon). This puts in doubt how much ICI was actually concerned with a 'competence-based approach' to technology strategy since it invested almost all of its resources in business-orientated research, reflecting what seems to be a much more 'traditional approach'.

Two years after the *Financial Times* report mentioned above ICI announced its separation into two independent companies, confirming that its present approach to technology strategy is in fact more traditional than that described by its research director. It also shows that, in spite of the fact that synergy is a credited strategic factor, there are several other aspects which can be even more important to the survival of large corporations (at least in the short term).[3]

In a more academic view, Twiss (1986, p. 59) discusses technology synergy as an important issue for take-overs: 'There are a number of reasons why one company should decide to take over or merge with another. Synergy and the matching of resources often play an important part in the industrial logic for take over. Technology can be a critical factor.'

In terms of long-term survival there are two important aspects to consider. The first is the search for a proper balance between mature and emergent technologies within the portfolio. The second is the potential effect of radical innovations upon industries' structures. Porter (1985, p. 172) illustrates succinctly the importance of the technology and technological change upon an industry structure:

Technology is also an important determinant of overall industry structure if the technology employed in a value activity becomes widespread. Technological change that is diffused can potentially affect each of the five competitive forces, and improve or erode industry attractiveness. Thus even if technology does not yield competitive advantage to any one firm, it may affect the profit potential of all firms. Conversely, technological change that improves a firm's competitive advantage may worsen structure as it is imitated. The potential effect of technological change on industry structure means that a firm cannot set technology strategy without considering the structural impacts.

The occurrence of innovations capable of threatening the whole structure of the chemical industry is quite improbable. Porter's text above is pertinent to more dynamic industries such as information technology and electronics. His comments on the influence of technology and its change upon the 'five competitive forces', however, are important to *any* industry mainly at the individual business level as discussed in the next section.

2.3 Technology and business strategy

Here, the objective is to discuss the possible effects of technology and technological change upon companies' competitive advantages, threats and opportunities. Porter (1985) conducted the analysis of the importance of technology in a firm's strategy from two broad points of view: (a) the competitive advantage; and (b) the industry structure.

2.3.1 Competitive advantage

In order to affect the competitive position of a firm, technology should play a significant role in determining the relative cost or 'differentiation uniqueness' of products or services. Porter listed some of the major roles technology can play through generic strategies. These roles are shown in Table 2.1 for each of his 'generic strategies' in terms of product and process innovations. In accordance with his thinking, any activity on the value chain involves a specific group of 'sub-technologies' or scientific disciplines,[4] and these 'sub-technologies' are dispersed throughout the production chain. In this way, it is not only scientific breakthroughs that can have a major influence on a firm's competitive advantage but also the association of incremental changes. Empirical data support this assertion, even for some leading innovators, as shown, for example, by Hollander (1965). In his study of the Du Pont rayon plants, he concluded that the minor innovations had a more important contribution (as a whole) to the company than the related radical breakthroughs. The importance of minor innovations is central in this work and will be subject to more investigation in later chapters.

2.3.2 Industry structure

In analysing the importance of technological change in the industry structure, Porter (1985) considers the relationship with suppliers, buyers, new entrants and substitutes.

In order to reduce buyers' bargaining power, he says, technology change can create differentiation and increase buyers' switching costs. He also raises the possibility of having a buyer backward-integration-threat enhanced by the buyer's own technological development. In relation to the new participants, the main consequences of technological change are related to entry and exit barriers, economies of scale, cost and uniqueness. These ideas, as well as those concerning relationship with suppliers and substitutes, will be fully explored in Chapter 3.

The importance of technological issues to business-level strategy appears to be more readily acceptable than would seem to be the case with corporate strategy. Support for this statement is provided by the fact that at present almost all companies have business-orientated research while the same cannot be said about corporate-orientated research. Another clear demonstration is given by the percentage of the R & D budget deployed in

Table 2.1 Technology and the generic strategies (adapted from Porter, 1985).

	Cost Leadership	Differentiation	Cost Focus	Differentiation Focus
Product Innovation	Product development to reduce product cost by lowering material content, promoting easier manufacture or to simplify logistics.	Product development to enhance product quality, features, deliverability or switching costs.	Product development to design in only enough performance for the target segment's needs.	Product design to meet the needs of a particular segment better than broadly targeted competitors.
Process Innovation	Learning curve process improvement to reduce material usage or lower labour input. Process development to enhance economies of scale.	Process development to support high tolerances, greater quality control, more reliable scheduling, faster response time to orders, and other dimensions that raise buyer value.	Process development to tune the value chain to a segment's needs in order to lower the cost of serving the segment.	Process development to tune the value chain to segment needs in order to raise buyer value.

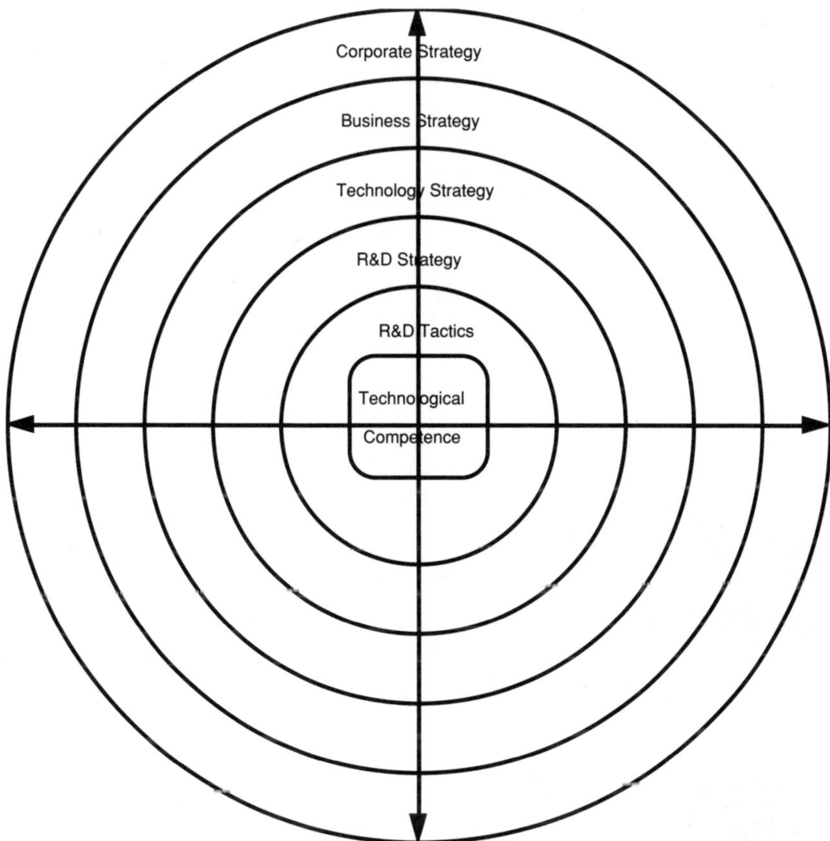

Figure 2.1   The strategic view of technology through different levels.

these two different types of research. Cookson (1991) says that traditionally the amount invested by 'science-based multinationals' in corporate projects is just around 10 per cent of the R & D budget.

In spite of these strong indications of a disseminated 'business orientation' on industrial research and development, Szakonyi (1990, p. 392), in surveying more than 170 American companies, concluded that less than 10 per cent of them had effective coordination between R & D and business planning. Similar results were obtained by Weil and Cangemi (1983, p. 33) from a survey of 55 companies where they found that 65 per cent of the R & D managers complained about the necessity to improve this relationship. Fusfeld (1978) and Harris *et al.* (1983, p. 30) also obtained similar results. They found that 86 per cent of the companies surveyed lacked an analytical approach to the question. This subject is among those discussed in the next section, where the role played by strategy in the main loci of the technology-generation process (the technology and R & D functional areas) is the main focus.

2.4 Strategy at the functional level

Most of the authors cited in the previous chapter recognize the idea of strategy at the functional level such as a financial strategy, a marketing strategy and a research and development strategy. Other authors used to call the same concepts 'functional tactics' – a definition which has not been adopted in this research (as described in the previous chapter).

Hofer and Schendel (1978, p. 29) made the following comments about 'functional area strategy':

> At the functional area level, the principal focus of strategy is on the maximization of resource productivity. Synergy and the development of distinctive competence, therefore, become the key strategy components, while scope drops sharply in importance. Here, synergy involves the coordination and integration of activities within a single function.
>
> While each of these types of strategy are distinct, they should all fit together to form a coherent and consistent whole for any particular organization, if the organization is to be successful over the long run. This requires that each level of the organization be constrained by each other level, which usually requires functional area strategy to be constrained by business strategy and it, in turn, to be constrained by corporate strategy.

In line with Hofer and Schendel's comments, this book is concerned with strategy, with special attention being given to the business level but also to moulding a broad view of the strategic management of technology as shown in Figure 2.1.

2.4.1 A strategy for technology

Following the concepts adopted in the previous chapter, technology strategy will be analysed here as a pattern through which companies or strategic business units (SBUs) approach their technological objectives. This general idea is not in conflict with other authors such as Porter, Bitondo and Frohman, Szakonyi and others. Conversely they can be combined and perhaps improved from the analysis of their weaknesses and strengths as shown below.

Porter (1987, p. 66) defined 'technology strategy' as a firm's approach to the development and use of technology. In his view it must address three broad issues:

  (i) – What technologies to develop;
 (ii) – The role of technology licensing; and
(iii) – Whether to seek technological leadership.

Bitondo and Frohman (1981, p. 21), list what they called the technology strategy-dimensions as:

  (i) – Positioning the R & D strategy from offensive to defensive;
 (ii) – Posturing R & D effort from technology invention to technology application;

(iii) – Product-line strategy ranging from eliminating products to new product development;

(iv) – Matching the needs of products requiring R & D from product to process innovation;

(v) – Maturity of technologies in their own life-cycle; and

(vi) – Strategy for acquiring technology.

Szakonyi (1990, p. 393) advocated a 'technology plan', consisting of decisions about which technologies a company should concentrate on and which not to support. In his view such a plan should be based on the assessment of:

(i) – R & D organization's capability;

(ii) – Immediate and strategic business needs that an R & D organization must serve;

(iii) – Future technical potential of various technologies; and

(iv) – Leverage that those technologies could provide in business applications.

Szakonyi's work was intended to provide guidance to R & D from business planning. In this way the main issues, he emphasized, from the technology plan are in the domain of the R & D activities (see next sub-section). Fusfeld (1978), (1988) and Clarke *et al.* (1989), to mention but a few, also used the concept of technology strategy as some type of link between R & D and the business environment. In the following paragraphs some of the main points will be discussed, as well as the main divergences within these works, leading to the definition of technology strategy that will be used in the later case-studies.

Porters's technology-strategy definition and issues (shown above) are very general and consequently comprehensive. As such they will be used as the starting point for discussion as follows:

(i)   To choose the 'technology to be developed' Porter suggests, as a first point, the need to consider the company's generic strategy (see Table 2.1). In doing so, a firm should concentrate on those technologies that have the greatest sustainable impact on cost or product differentiation. Bitondo and Frohman based their work on a group of five matrices derived from the Boston Consulting Group (the three which are directly related to technology are shown in Figure 3.7). Having recognized the existence of a technology's own life-cycle as one of the problem's 'dimensions', their methodology seems unable to provide a clear distinction between that cycle and that of the product. The concept of a technology life-cycle is disregarded in the methodology, which at the same time is heavily based on product cycles. Also with regard to what concerns technology life-cycles, Porter (p. 67) states: 'It is only a particular combination of sub-technologies that can be assumed to be mature, not sub-technologies themselves.' His observation, although true, must be interpreted with considerable care. The high degree of integration of technologies in the chemical industry and its capital intensiveness (particularly in the petrochemical sector) makes quite usual the obsolescence of an entire process owing to the maturity of just one of the technologies or unitary operations which it employs.

In terms of technological leverage for business, however, Szakonyi's view coincides with Porter's but goes further by taking into consideration the future technical potential of technologies (implicitly differentiating products and technologies). Methodologically, Szakonyi also goes further by suggesting feedbacks from business planning such as product-line strategies (a three-year forecast describing the necessary characteristics of present products) and business strategies (with a forecast of new products for the next ten years).

(ii) In relation to 'technology licensing', Porter investigated several issues. The most important seem to be related to:
- The inability to exploit the technology (the main causes of which could be: lack of resources, harvesting of the SBU or competitors who are already too entrenched);
- Rapidly standardizing the technology (which he illustrates with the competition between Betamax and VHS formats in video cassettes);
- Poor industry structure (where licensing can be a good deal for those industries with very powerful suppliers or buyers, for example);
- Creating good competitors (sometimes they can stimulate demand, blocking the entry of more powerful competitors).

Porter's analysis only covers the view of the licenser. Nevertheless, the licensee can also have a broad technological strategy, of which buying a license may be only one issue. In complementing Porter's view, it could be suggested that the main reasons to look for a technology licenser could be:
- Lack of technical skills, financial support or time availability;
- Patenting or standardization of down-stream industries; and other common reasons such as commercial agreements.

On the other hand, the main reasons not to search for a technology licenser can be usually linked to:
- The existence of in-house technical and financial capability or availability of attractive external hiring possibilities.
- The absence of an available licenser – an increasingly important factor for in-house development, particularly in the sector of speciality chemicals.

As suggested by Bitondo and Frohman's work, it seems interesting to exchange Porter's 'licensing-role' issue for a more comprehensive analysis which allows the consideration of other types of technology acquisition such as contracting, joint ventures and co-operative research as well as the possibility of company strategies involving their role as licensees.

(iii) In relation to 'technological leadership', Porter classified the technological strategies as 'leader', 'followers' and 'disregards'. Such

Table 2.2 Freeman's classification of strategies (adapted from Freeman, 1982).

	Offensive	Defensive	Imitative	Dependent	Traditional	Opportunist
Fundamental research	4	2	1	1	1	1
Applied research	5	3	2	1	1	1
Experimental development	5	5	3	2	1	1
Design engineering	5	5	4	3	1	1
Production engineering Quality control	4	4	5	5	5	1
Technical services	5	3	2	1	1	1
Patents	5	4	2	1	1	1
Scientific and technical information	4	5	5	3	1	5
Education and training	5	4	3	3	1	1
Long-range forecasting and product planning	5	4	3	2	1	5

Range 1 - 5 indicates weak (or non-existent) to very strong

classification was used before in some more comprehensive and deeper taxonomies created by Freeman (1982) and Twiss (1986).

Freeman, while recognizing that any classification of strategies by type necessarily involves some arbitrary decisions, observed and described six types of strategy: offensive, defensive, imitative, dependent, traditional and opportunist. Stating that they can be seen in a continuous spectrum of options, he found that 'some companies recognisably follow one or other of these strategies'. Table 2.2 shows the emphasis each of those stereotypes identified by Freeman put on fundamental versus applied research; experimental development versus design engineering and some other characteristics.

While Freeman called these types 'innovation strategies', Twiss defined his own taxonomy as R & D strategies. They are: 'offensive', 'defensive', 'licensing', 'interstitial', 'market creation', 'maverick', 'acquisition-people' and 'acquisition-companies'. Twiss saw the strategy selection as an iterative process of analysis of environment, capability, portfolio and individual projects. He not only recognized the spectrum of possibilities but also the use of a portfolio of strategies: 'For example the choice will not be between an "offensive strategy" or a "defensive strategy" but the proportions of the available resources to be allocated to each of them.'

This type of strategy taxonomy was commented on by Coombs (1990):

… these stereo-types, whilst having quite definite implications for R & D resource allocation and style, did not reduce to a description of only the R & D function and its management. On the contrary, the strategies were seen as inclusive of appropriately resourced and directed marketing efforts and production efforts. Thus strategies are, in a sense, innovation-sensitive aspects of the overall business strategy of the firm.

Coombs's comments are extremely pertinent in relation to Twiss's taxonomy. However, Freeman's is actually defined as a classification of 'innovation strategies' which can be fully incorporated into a broader concept of 'technology strategy' (which may involve even the market efforts mentioned by Coombs).

As observed by Coombs et al. (1987), it is useful to isolate these elements in the search for 'a discrete domain of R & D strategy or technology strategy'. This search will be attempted throughout this book, recognizing, however, that they are not necessarily discrete but can at least be divided into two different domains: 'the technology strategy' and, as part of its content, the 'R & D strategy' (focused on in the next sub-section). That assumption of concentric strategies is in accordance with Hofer and Schendel's and Porter's views (as described above) and was illustrated in Figure 2.1.

It is interesting to note that in citing the taxonomies of Twiss, Sethi et al. (1985) and Steele (1975), Szakonyi commented that regardless of the classification used, the important point is for the managers to know which strategy the business will follow, allowing them to be aware of which contributions can be made in-house and which should be obtained from outside. Such observation agrees with and complements the main conclusions from the work of Weil and Cangemi (1983) in which they show that the main problems observed in the linkage of long-range R & D and strategic planning are: the mismatch of time horizons between business planners and researchers; the lack of knowledge about corporate goals by researchers and the difficulty in making long-range forecasts of markets and technologies.

By considering all these observations, comments and criticisms, it seems quite natural to analyse a firm's technology strategy through a concept involving the following main issues:

(i) The approach to be used:
- Traditional, where the technologies to be developed are defined from the business strategy (as derived from Porter's and Szakonyi's views);
- Competence-based, where the existing technological competence within the firm is directed towards new products and markets (as derived from Fusfeld's work).

(ii) Definition of the role of external technology generators from the internal capability audit (as derived from the work of Porter, Szakonyi, Twiss and others).

(iii) The share of the effort to be put in fundamental versus applied research; experimental development versus design engineering and other options in the spectrum as observed by Freeman but also considering that different styles can be deployed to the different projects within the portfolio (as derived from Twiss).

From the consideration of all these issues the concept of business strategy will here be defined as:

A pattern of resources and competencies deployed through environmental interactions to achieve a business goal or objective.

In order to allow these issues to be consistent with the concept of strategy defined in the last chapter and with the three points depicted above, the concept of technology strategy will be used in this research in accordance with the following definition:

A pattern of technico-scientific resources and competencies deployed to achieve a technological goal or objective defined from the analysis of the business's competitive forces.

2.4.2 A strategy for R & D

The research and development function is the core of industrial innovation. Recognition of this factor dates back to the 1870s when the present giant German companies were world pioneers in formal industrial research. At that time, the process of innovation was extremely dependent upon 'entrepreneurship'. Freeman (1982) clearly shows that the maturing of the scientific nature of technology made this dependence move towards the professionalization of the R & D activity (see Chapter 4).

Coombs (1990) in analysing some highly professionalized R & D departments of big industrial enterprises, observed:

The need for managers to identify technologies and products separately and give strategic attention to both portfolios appears to be an increasing one. Thus it can be argued that the rate of increase of technological diversity at firm level appears to be greater than the rate of increase in product diversity.

This observation seems to be in some opposition to what had been written in 1983 by Linn: 'The goals of R & D are the business goals of the firm and are not measurable solely by technical success and failures. *R & D effort unrelated to corporate goals and objectives is a waste of time and money*' (author's own emphasis). Coombs's observation about the dichotomy between products and technologies is a relevant one. It can help to avoid the potential danger of mismatch between, for example, the maturity of technologies with that of products or markets. Linn's view, however, if followed without considering the different nature of products and technologies, seems to lead directly to that mismatch. This discussion returns to the two approaches for technology strategy (the traditional and the competence-based). A direct link between R & D and business, disregarding the different (although complementary) natures of technologies and products, can be associated with the traditional approach. The exploitation of those differences, on the other hand, makes easier the development of new products or markets to explore the firm's existing competence.

The notion of R & D strategy can be closely related to the business strategy but also should have some autonomy, since it involves a defined managerial sub-structure and is the core of the innovation process. The degree of linkage between the business strategies and the R & D activity is then one of the main issues of technology strategy.[5] Shota and Bloack (1987) showed the adoption by big chemical companies of a reasonable approach to the problem. Such an approach involves the creation of 'R & D – business linked' departments as well as those 'R & D – corporate linked' (also illustrated by Coombs). This solution is intended to keep the research effort business-orientated, at the same time that it allows the existence of long-term independent research as a source of innovation (as well as a motivation for researchers). At the theoretical level, this approach was also described by Wilkinson (1987). Whichever approach is adopted, it seems fundamental to have clearly defined the R & D mission which necessarily involves the responsibility of creating new business opportunities and/or competitive advantages in the present businesses or, at least, to keep pace with competitors' advantages.

Twiss (1986) made some comparisons between R & D and corporate strategies where he shows, as one of the main issues of the R & D strategy, the allocation of resources among projects. The importance of this issue is also recognized by Coombs *et al.* (1987; Coombs 1990); Freeman (1982) and Liberatore and Titus (1983), the last of whom made an extensive survey of 32 companies conducting R & D strategic planning and found results in support of this observation about resources allocation. At the same time they showed several other important and common issues of such planning which are depicted in Table 2.3.

Certain results shown in that table seem to be functions of a 'technology strategy' instead of an R & D one (according to the definitions here adopted), although the authors do not use such a concept in their paper.[6]

Table 2.3 Main issues of R & D strategic planning (adapted from Liberatore and Titus, 1983).

Functions / activity	Percent
Supports new ventures / products for growth	97
Allocates project / program budgets	91
Identifies R&D resources gaps	88
Integrates business strategy / plans	75
Performs competitive technology evaluations	44
Incorporates formalised technological forecasting	9

2.5 Conclusion

While almost 40 years have passed since strategy became a subject for attention and study in the business as well as in the academic worlds, it has been no more than 15 years since its linkage with technology appeared as a serious subject. The majority of the studies in this area have been conducted in the last decade and only in the most recent years has the analysis achieved a considerable level of sophistication. As a consequence, most of the conclusions which can be drawn from a bibliographic review are only hypotheses and assumptions.

At the level of corporate strategy, it can be seen that the perception of technology as a strategic factor has increased. It can also be concluded that one of the main issues from this strategic view of technology is the concept of synergy, corroborating with the dissemination of the competence-based approach.

At the business and functional levels the perception of technology as a strategic subject and the use of techniques and tools seems more in evidence. This poses the main question as to how much are the traditional and the competence-based approaches spread throughout the industry. Other very fundamental questions are: how does the technology strategy process of genesis occur? Are there reliable techniques and tools in use or is it merely still a question of managerial good sense?

These questions will be analysed more deeply in the next chapter, where the main subjects include the tools and techniques described in the literature.

Notes

1. If strictly following the sequence, we could identify a third locus, the contribution of technology to the functional strategies, leaving the 'technology strategy' as a fourth. Technology actually can have a very important role in several functional strategies such as production, quality control and even finance. However, this research is only concerned with technology management along the hypothetical line that links R & D activities to (and from) business and corporate strategies.

2. Later in the same paper Fusfeld suggests the construction of a chart of technologies against a products portfolio. The aim of this chart is to aid in the development of an overall technology strategy from and to the business portfolio.

3. A different type of synergetic effect that can be generated by technology at the corporate level is linked to the marketing functional area. Morita (1987) illustrated such an effect (through a non-academic view) in the well-known book *Made in Japan*. He shows the strong link his company strived to build between its logo and the concept of technology and quality. This link now seems identifiable to customers, justifying the investment made in technology and on its association with the image of that firm.

4. While Porter sees technology as a group of 'sub-technologies', other authors such as Clarke *et al.* (1989), for example, refer to the whole as a 'cluster', composed of different technologies.

   Considering the scope of this book, it seems that the latter terminology is more appropriate. The chemical industry has become increasingly more dependent on a scientific base (as clearly shown by Freeman, 1982). Each unit operation such as distillation, extraction or reaction, for example, involves different technologies in their design, operation and control. Each of these technologies involves many different scientific disciplines. A digital-control system, for example, necessarily involves a myriad of electronic hardware, software, processes and simulation disciplines. Similarly, we can divide each production process into several clusters of technology, prolonging the division process as long as required. This division process could probably end in basic disciplines such as chemistry, physics and mathematics. The conclusion is that the degree to which it is interesting to subdivide the component technologies of a production process is necessarily determined by the objectives of such a division.

5. This degree of synchronism between R & D and business is within the main issues of a 'technology strategy' as described above. The raising of such an issue, however, is a direct consequence of the analysis of the R & D strategy itself as shown in this discussion.

6. There appears to be a tendency in the literature to accept this dichotomy between technology and R & D strategies. This tendency, however, was not explored at the time of Liberatore and Titus's research.

3 Technology-strategy formulation: aids, methodology and framework of analysis

3.1 Introduction

This chapter is intended not only to describe the methodology used in this research, but also to describe some broader aspects of its framework of analysis. Between these methodological aspects a description of the research's scope and of some of the concepts adopted will be given.

The last chapter was dedicated entirely to a theoretical discussion of the relationship between the concepts of strategy and technology. As stated before, most of the literature reviewed there comes from the area of management. This chapter will review some of the main works in the literature which attempt to utilize that theoretical contribution from the 'management school' in terms of tools and techniques (where 'innovation literature' has a major influence). The first step will be to divide the technology-strategy study into three analytical areas, following the work of Mulder and Vergragt (1990). Using their model of three different levels of analysis (which together form a framework for the approach to the individual companies), a short review of the techniques and tools available in the literature, and which fit into each of the different levels, will be presented.

3.2 The collective approach

As described in Chapter 1, there are hundreds of different definitions of the term 'strategy', although all of them are in some way related to survival. This observation alone accounts for the great difficulties in obtaining 'strategic information' from the companies studied. It also suggests that research based upon a survey may be too superficial to conduct an analysis about business strategy and technology strategy. In order to ensure the attainment of relevant and reliable information, it is essential to build up a relatively intimate relationship with the companies under analysis. Although the use of surveys can generate a much larger amount of data, the openness the subject requires is only compatible with research conducted through in-depth case-studies.

While the necessary level of information disclosure is only compatible with case-studies, this approach is not sufficient to guarantee obtaining the required information. In order to optimize the probability of getting this, it seems very important to ensure the following:

– To contact the right people in each company. This means personnel working with business strategy and people working with technology and technology strategy. In other words, medium to high levels of management in each of these areas.
– To be able to discuss technology and business strategies at the same level as the people contacted in each company. This means an ability not only fully to understand the subject and its semantics but also the ability to propose alternatives and to question decisions. Such an ability can only be developed through the study of the different products involved in the research.

For the above reasons, each case-study involves the analysis of production processes, applications and markets.

3.3 Scope of research and associated concepts

The type of data required (as described above) involves different disciplines and can lead to severe operational difficulties. In order to assure its feasibility, it is useful to conduct the research upon a supply chain (or building block) instead of using a 'horizontal sampling'. Then, the knowledge obtained from one level of the chain can be used in the understanding of the following level. A further justification for such an approach is the existence of networks of personal knowledge in such chains which facilitate access to the right individual in the different companies.

To secure the legitimacy of the conclusions drawn from this research, the building block was chosen in consideration of its economic importance as well as its representativeness in terms of marketing and technology. In order to justify the adoption of a supply chain in an examination of the petrochemical industry, it is necessary to review some of the characteristics of the different types of chemical products. Section 3.3.1 below describes some of the attributes of the different groups as well as some other concepts which will be necessary in other parts of this book.

3.3.1 Associated concepts

This sub-section presents the definitions to be used in this work in relation to sometimes controversial concepts such as 'chemical-processing industries', 'petrochemicals', 'chemical commodities', 'pseudo-commodities', 'specialities', 'fine chemicals' and 'innovation'. While the first six concepts are directly connected with this chapter, 'innovation' is much more relevant to the following chapters. As was also the case with the terminology presented in Chapters 1 and 2, definitions given here are intended only to indicate interpretations which have been controversial. It is intended that these definitions should be seen as generic guidelines and not as rigid borders between concepts.

The first problem in defining the research's scope as 'only one supply chain' is that there is no one capable of representing 'chemical-processing industries' (CPI) in their entirety. This first definition (CPI) includes all industries in which a chemical reaction has a major role to play. Consequently, it involves some very distinct industries such as oil refineries, wineries, paper and steel mills. A more useful widely accepted concept is given by the American Standard Industrial Classification (SIC) groups 28 and 29. The first is defined as including 'establishments producing basic chemicals, and establishments manufacturing products by predominantly chemical processes...' The second 'includes establishments primarily engaged in petroleum refining, manufacturing paving and roofing materials, and compounding lubricating oils from purchased materials'. While the firms in group 29 are mainly the so-called 'oil-companies', those in group 28 are more typically truly chemical companies. In considering the importance of the different sub-groups within group 28 for society as a whole, it can be seen that sub-groups 286 (industrial organic chemicals) and 282 (plastic materials and synthetic resins, synthetic rubber, synthetic and other man-made fibres except glass) are certainly among the most important. Apart from their economic importance, the two sub-groups are very interdependent but still have some technological and managerial similarities to group 29, thus making an ideal group in which to concentrate the search for a supply chain.

A further concept which can be extracted directly from the definition of sub-groups 282 and 286 is that of 'petrochemicals'. The literature presents several definitions for petrochemicals such as:

'The petrochemical industry may be generally defined as comprising those areas of chemical manufacturing which use raw materials extracted wholly or largely from petroleum or natural gas' (Riegel, 1974, p 426);

'More recently the term "Petrochemicals" has assumed a wider meaning, namely chemicals made by the large-scale petroleum chemical industry' (Stern, 1971).

Some definitions are not just wider but also 'deeper' as they take into consideration a 'vertical' component to include intermediate and end-products derived from subsequent chemical processing of basic petrochemicals. This second type of definition can be seen, for example, in (Hampel and Hawley, 1976, p. 207): 'A chemical intermediate obtained directly from petroleum or natural gas... or an end-product derived by subsequent chemical processing.' This last definition fits quite well with the idea of a supply-chain probing.

In terms of marketing there are four main groups of chemical products:

Table 3.1 The four groups of chemicals.

Production volume	Undifferentiated	Differentiated
High	Commodities, e.g. sulphuric acid	Pseudo-commodities e.g. polypropylene
Low	Fine-chemicals, e.g. ascorbic acid	Speciality chemicals e.g. pesticides

commodities, pseudo-commodities, specialities and fine-chemicals. In order to understand this classification, it is first necessary to classify chemical products into two broader groups: differentiated and undifferentiated. While the first group involves chemicals with characteristics intrinsic to their producers and usually priced in accordance with some type of performance index, the latter can be obtained from several different sources and still have the same specifications and composition.

If a chemical product is produced in large volumes and is undifferentiated it is classified as a commodity chemical; sulphuric acid is the most usual example. Pseudo-commodities are those products which, although being produced in high volumes (typically several million tonnes per year world-wide) are still slightly differentiated. Typical examples are polymers such as polyethylene. Although any polyethylene basically has the same type of composition, the average size and distribution of its molecules can change quite dramatically from one producer to another. Such differences will result in distinct properties. As a consequence its producers will be able to sell their products in accordance with their performance. In contrast with chemical commodity producers, pseudo-commodity producers often offer technical assistance and advertise their products. Fine-chemicals are produced in much smaller volumes and are usually undifferentiated. A good example is ascorbic acid (or Vitamin C). Specialities are also produced on a smaller scale than commodities and are highly differentiated when compared to fine chemicals. Good examples are formulated pesticides and industrial explosives. Such products are often sold to end-users and advertised. A good description of several characteristics of the four groups can be found in Wey *et al* (1979). Table 3.1 illustrates the differences among the four groups.

Although the definitions above will be necessary for a full understanding of this work, the four groups will be regarded in most of this book as just two major groups: commodities (involving both high-volume groups) and specialities (involving fine-chemical and specialities themselves). Such a simplification will not only make some of the discussion easier but, more

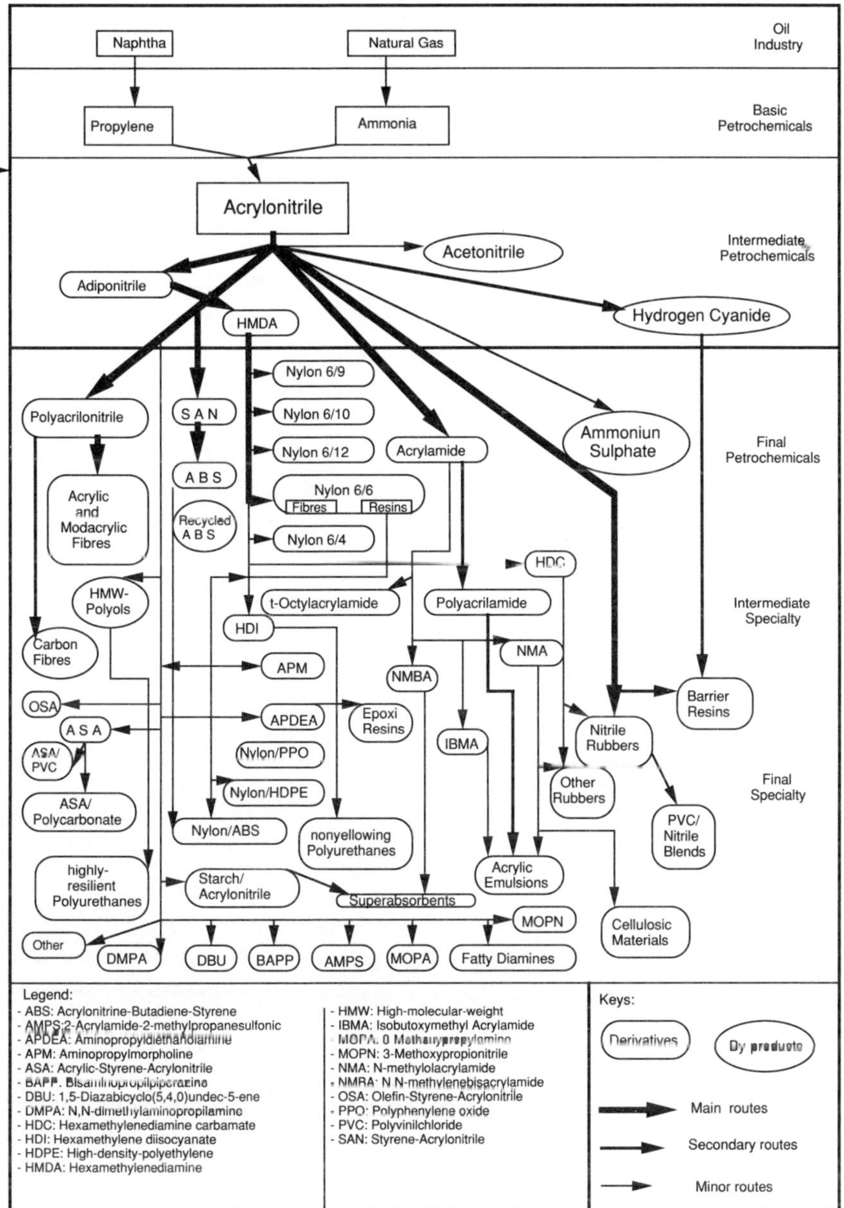

Figure 3.1   Acrylonitrile's supply chain.

importantly, also follows the nomenclature usually found in the 'market place'.

Another essential definition for this research is that of 'innovation'. The concept of innovation is generally recognized (including in this book) as the first introduction in the market of a new product or process. This interpretation is attributed to Schumpeter. The broad range of technical changes, from radical breakthroughs to incremental changes, analysed in this research, however, demand some complementary ideas such as the definition of 'internal innovation'. This concept differs from Schumpeter's broader concept of innovation only in terms of scope. While the former is concerned with the first application of a product or process in the economy as a whole, the internal innovation can be defined as 'the first introduction in the market of a product or process by a specific company or SBU'.[1] While this second interpretation has a restrict nature the first is understood to be an 'absolute definition'.[2]

3.3.2 The acrylonitrile building block

Acrylonitrile is a petrochemical intermediate commodity with five main derivatives (acrylic fibres, adiponitrile-nylon, ABS/SAN resins, acrylamide and nitrilic rubbers) and dozens of minor ones as illustrated in Figure 3.1. Its supply chain involves commodities, pseudo-commodities, specialities and even some fine-chemicals, representing an interesting collection of products.

Some other reasons for choosing acrylonitrile and its down-stream industries as a focus for this research are:

1 – Great economic representativeness;
2 – High number of products and producers in the United Kingdom;
3 – A 'technological monopoly' and consequently competitive advantage of the basic product manufacturer in the first stage of the supply chain associated with intense technological and commercial competition in several sub-sectors from the second stage in the chain;
4 – There are products in all stages of the life-cycle.

The present world production of acrylonitrile is around 4.35 million tonnes per year. North America is the largest producer region, responsible for some 1.4 million tonnes, followed by Western Europe (1.1 million), Japan (0.6), the rest of Asia and the Far East (0.5 million), Eastern Europe (0.4 million), Latin America (0.25) and other regions (0.1). Consumption is largest in Western Europe followed by Asia/Far East, Japan, North America (a very large net exporter), Eastern Europe, Latin America and other regions. The acrylonitrile market involves nearly US\$3 billion each year, a value certainly widely surpassed by its derivatives.

Acrylic fibres are long-chain synthetic polymers composed of at least 85 per cent acrylonitrile units. Modacrylic fibres are a sub-classification with a 35–85 mass percentage of acrylonitrile. They are used mainly in apparel and home furnishings such as sweaters, single- and double-knit jersey, hosiery, pile, fleece goods, carpets, blankets, curtains and drapes. World production of acrylic fibres reached 130,000 tonnes in 1960, 400,000 in 1965, 1 million in 1970 and 1.4 million in 1975 (Kirk Othmer, 1975). At that time its market was already more than US\$ 2 billion per year; presently its world production is around 2.4 million tonnes.

ABS/SAN resins contain up to 35 per cent of acrylonitrile and, after acrylic fibres, are expected to remain as the major users of this petrochemical. Because of these derivatives, acrylonitrile is today one of the more important building blocks of the plastics industry. The largest use of SAN resins is in the production of ABS resins which are obtained by various combinations of the former with grafted rubber. The two resins together represent more than 5 per cent of the plastics market. As early as 1975, ABS production in the United States was around 625,000 tonnes. The versatility obtained by the infinitely different blends results in a broad application – from refrigerator shelves and brush handles to lenses and car parts (which are the main final use of ABS resins). Present world production is about 2.7 million tonnes.

Adiponitrile is manufactured mainly for use as an intermediate for hexamethylenediamine, which is the main ingredient for nylon 6/6 production. The latter is produced as fibre for the textile industry and as resin with many other industrial applications. Reacting with different diacids, HMDA will also produce nylon 6/9, 6/10 and 6/12. Its world production is around 0.43 million tonnes.

Acrylamide has a world-wide production of roughly 200,000 tonnes per year, worth approximately US$ 200 million . Its main uses are water-soluble polymers applied in water treatment, enhanced oil recovery, papermaking and as leather treatment aids. It is also used as an additive in alcohol, sugar-production processes and the textile industry.

Nitrile rubbers are extensively used in process industries because of their resistance to chemicals, oils, solvents, heat and abrasion. Applications that do not require high solvent resistance use lower acrylonitrile grades which still maintain good low-temperature flexibility. Many products such as fuel hoses and cell liners, seals and gaskets employ nitrile elastomers. The world's present production is approximately 0.25 million tonnes.

3.4 The individual approach

Mulder and Vergragt (1990) analysed the emergence of aromatic polyamide-fibres technology in relation to the strategies of some of the leading industrial companies in the field of synthetic fibres. In that work they successfully tried not only to analyse the strategies employed but also to show that it is possible and fruitful to distinguish three different levels of strategic analysis: macro-, meso- and micro-levels.

3.4.1 Macro-level

The macro-level theoretical basis comes from economic theory, mainly from the work of Nelson and Winter (1977), Abernathy and Clark (1985) and Dosi (1982).

Nelson and Winter studied technological progress by introducing the concept of 'technological regime'. This concept dealt with potential, constraints and unexploited opportunities. They state that: '...this sense focuses the attention of engineers on certain directions in which progress is possible...' (p.57). The combination of all these factors is a technological momentum that results in a 'natural trajectory'.[3]

Dosi introduced the concept of 'technological paradigm'. The concept involves continuous and discontinuous processes of technological change. The continuous changes 'are often related to progress along a "technological trajectory" defined by a technological paradigm, while discontinuities are associated with the emergence of a new paradigm' (p. 147). He relates this concept to that of technological trajectory by suggesting that these trajectories are a clustering of possible technological directions whose outer boundaries are defined by the nature of the paradigm itself.

Abernathy and Clark introduced the concept of 'dominant design' and used it to describe the periods of minor (regular) change. According to their analysis there are four different types of innovations:

– 'Architectural innovations' are more radical. This type of innovation disrupts the existing competencies in the industry and even the existent linkages with markets. Such innovations sometimes create a completely new market and constitute a dominant design.
– 'Regular innovations' follow a dominant design. However, they can generate successive improvements, entrenching the existent competencies and the linkages between producers and markets.
– 'Niche-creation innovations' reinforce existing competencies, but disrupt market linkages as they create specific niches for the innovator.
– 'Revolutionary innovations' disrupt existing competencies; nevertheless, they do not change the existing linkages with markets.

The three works cited are complementary and comprehensive, providing a basis for an economic innovation theory. Mulder and Vergragt used these theories for their 'macro-level analysis', stating (p. 248) that: '... all of these macro approaches cannot account for the different strategies of firms within the same field of technology and the different technologies they develop. They can only account for the part of the trajectory which all companies have in common...' In the authors' view, this level of analysis accounts not only for technological developments but for a cultural one as well.[4]

On page 247 they described the 'natural trajectory of synthetic fibres': 'New synthetic fibres could replace natural fibres like wool, cotton and silk and semi-natural rayon fibres by improved properties... A great number of new polymers could be developed which could be spun into fibres by varying existing spinning technologies...' This trajectory was a dominant one for the entire synthetic fibre sector for several decades. Addressing the sub-sector of polyamide fibres they wrote: 'In this history we can recognize a technology trajectory. Methodologically, one used the solution condensation method to create aromatic polyamides; and existing fibre spinning technology to spin fibres out of this polyamides.'

The technology trajectory in each of the industries down-stream from acrylonitrile will be examined to test Mulder and Vergragt's belief in such trajectories as general guidelines to business and technological strategies for all companies in each of those sectors.

3.4.2 Meso-level

The second level of analysis is where the interactions and competition among different companies and their strategies occur. Here it is possible to analyse the opportunities provided from the technological and cultural trajectories to the participants. Mulder and Vergragt found that this trajectory provides some opportunities which are circumstantial to the companies, their markets and socio-political environments. They emphasized that at this level the main considerations could be 'why firms enter a new trajectory?' and 'how they react to the choices other companies made' (in their work, Mulder and Vergragt explained why Du Pont, Bayer, Monsanto and AKZO entered the aromatic polyamide-fibre sector through different ways).

Porter (1980, 1985) provided a useful model (discussed in Chapter 1) for the analysis of business strategies. Based upon his work, Wilkinson (1987) created a systematic approach for using business strategies as a source of ideas for research and development projects. The models illustrated in Figures 3.2, 3.3 and 3.4 are derived from Wilkinson's work and adapted to the scope of this book (the chemical industry). Figure 3.2 shows the main suggested R & D counterparts to the threats observed in the strategic analysis of 'on-going flow of products'. It is necessary to emphasize that each factor (for example, 'few suppliers') can lead to several counterparts (for example, 'modify process to utilize alternative raw-materials', 'modify process to reduce switching costs', 'process development towards back-integration', etc.).

Figure 3.3 shows the same type of analyses, but this time within the 'strategic dimension' of new participants. Here the R & D counterparts are suggested in order to elevate entry barriers and to lower the exit barriers. (The smaller figure in the right lower corner of that diagram shows the suggested movement from fragile positions to stronger ones.)

One important observation must now be made in relation to the threat of substitutes. As, in principle, these substitutes are unknown, it may be impossible to create entry barriers. Nevertheless, the lowering of the exit barriers can be seen as a feasible counterpart even for unknown substitutes.

In analysing the industrial competitors, Wilkinson considered three main aspects: the maturity of technology, the portfolio of activities and the breadth of activities, as shown pictorially in Figure 3.4. In relation to the breadth of activities, Wilkinson states that R & D has a clear role to play in seeking to introduce product differentiation (to a whole industry or to a particular sector), or in giving the company a cost advantage in process know-how.

In considering the portfolio of activities, Wilkinson considers the use of some 'technological matrixes',[5] such as McKinsey's and Booz Allen's tech-

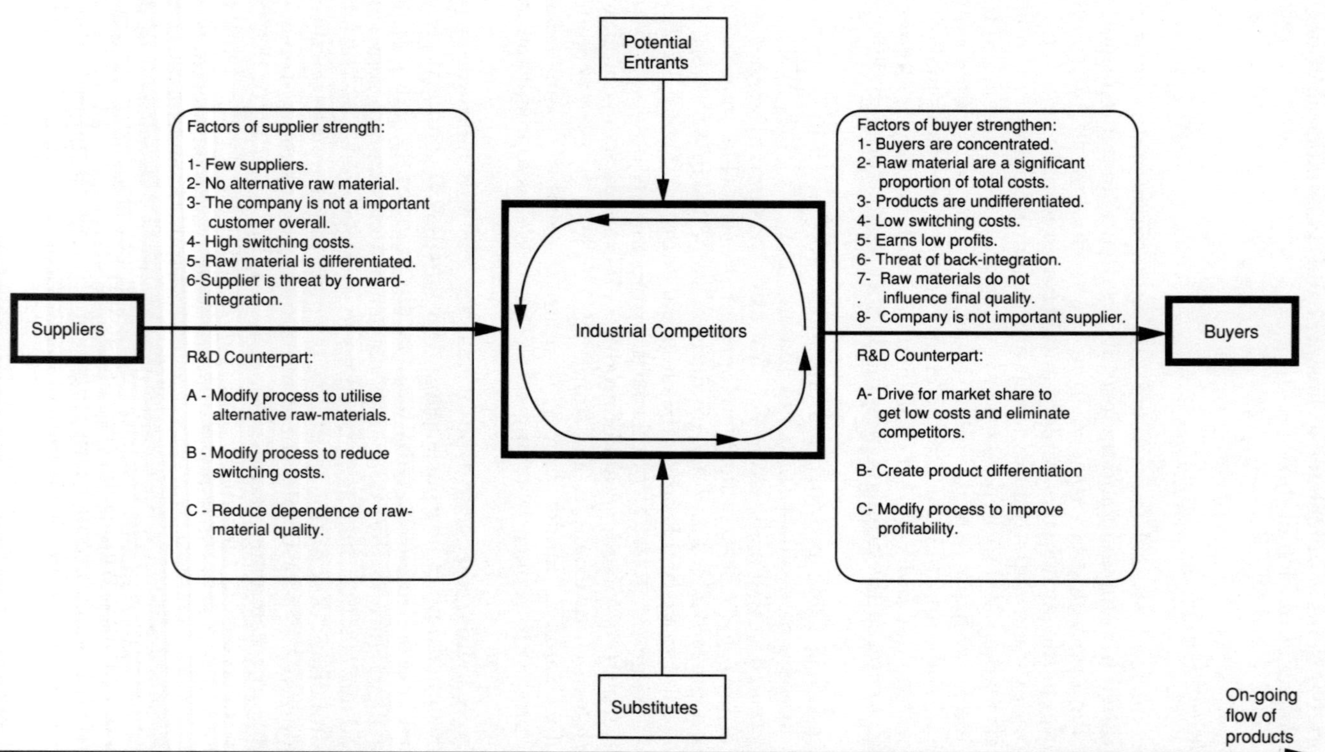

Figure 3.2 R & D project generation from threats from 'on-going flow of products dimension'.

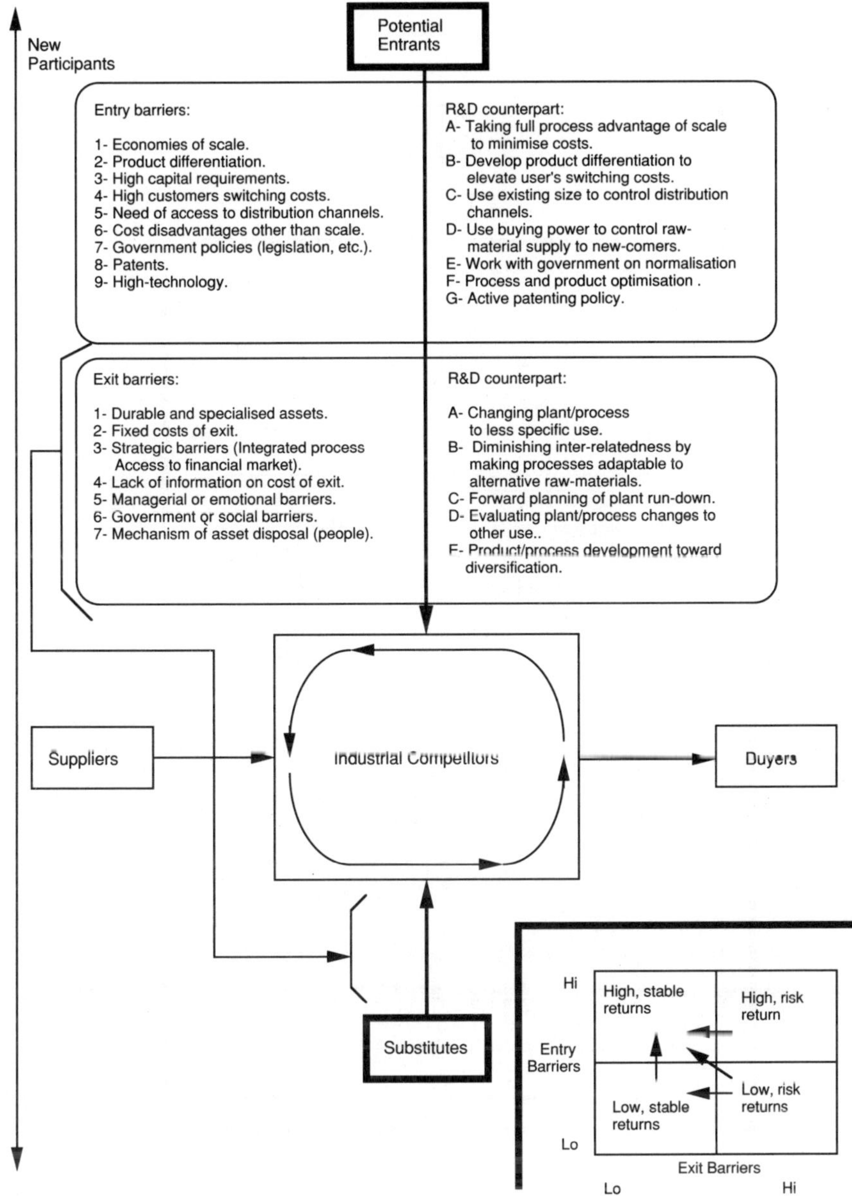

Figure 3.3 Project generation from threats from 'new participants dimension'.

nical matrices (see Figure 3.4). The use of this type of matrix was later
observed by Coombs (1990) in a giant chemical company and will be
discussed below. The use of those matrices may lead, however, to a potential mismatch between the maturity of technologies and products/markets,
requiring some care in their use and interpretation. This issue is one of the

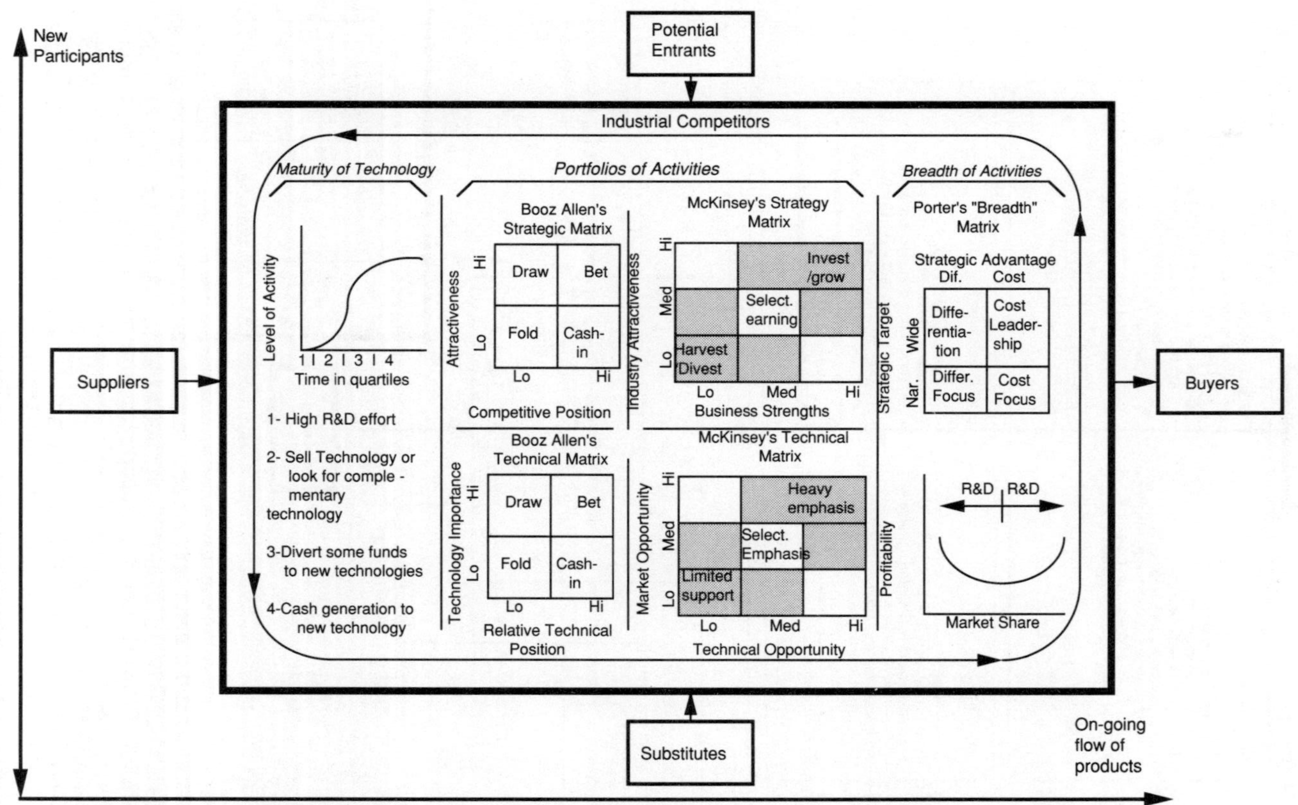

Figure 3.4 Pictorial view of Wilkinsons' technology-strategy generation model on the 'industrial competitors dimension'.

main subjects in some of the following chapters and in the field research. Care should also be applied to the third aspect considered by Wilkinson: the maturity of technology. This aspect was analysed by Wilkinson (1983) and is also summarised within Figure 3.4 where the straight connection, suggested by that author, between investment in R & D and maturity of the technology is indicated. This link can be compared with that made by Porter (1987) (as previously discussed), where he states that only certain combinations of sub-technologies can be considered mature.

3.4.3 Micro-level

The micro-level is where the analysis of the genesis of each company strategy is conducted. It has its roots in business administration and in the sociology of science and technology in accordance with Mulder and Vergragt. From the micro-level it is possible to analyse how strategy formulation and implementation within a firm takes place.

Within this third level of analysis some studies can be found in the literature, aimed at developing procedures for developing strategies. Some examples of these studies are given in Figures 3.5 and 3.6. These figures were developed from the study of the works of Sethi *et al.* (1985) and Bemelmans (1979). Both of these are extremely complex and seem quite mechanistic in their conception. Some parts of these works are, however, often found in the literature such as:

- Technological forecasting;[6] discussed by Twiss (1986), Harris *et al.* (1983), Szakonyi (1990) and others.
- Technology audit; discussed by Fusfeld (1987), Harris *et al.* (1983), Clarke *et al.* (1989), Twiss (1986), implicitly by Szakonyi (1990), Bitondo and Frohman (1981).
- Project portfolio; discussed by Harris *et al.* (1983), Coombs and Richards (1990), Twiss (1986), Wilkinson (1987), Szakonyi (1990), Bitondo and Frohman (1981) and others.

For the project portfolio, a frequently used tool of analysis (in the literature) is the so-called technology or technological matrix. Some of them (Booz Allen's and McKinsey's) have already been illustrated in the previous sub-section. Others of importance are shown in Figures 3.7 and 3.8. The former was extracted from Bitondo and Frohman (1981, p. 20 and 21), while the latter is from Coombs and Richards (1990).

Bitondo and Frohman matrices are aimed at:

(a) 'Positioning the R & D strategy from defensive to offensive, depending on the market attractiveness and business position.'

(b) 'Posturing the R & D effort within the spectrum of state of the art research (technology invention) to prototype development (minor application of existing technology), depending on product line need and market share strategy.'

(c) 'Product line strategy ranging from eliminating products to developing new products to new specifications.'

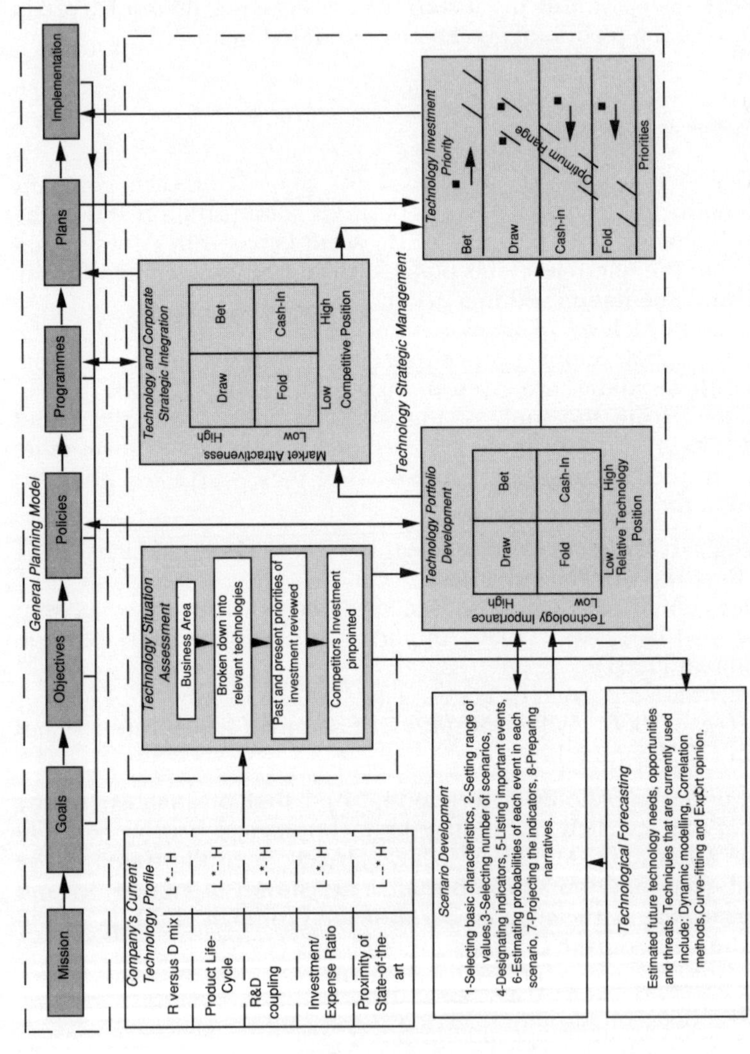

Figure 3.5 Model of technological-strategy's genesis (adapted from Sethi *et al.*).

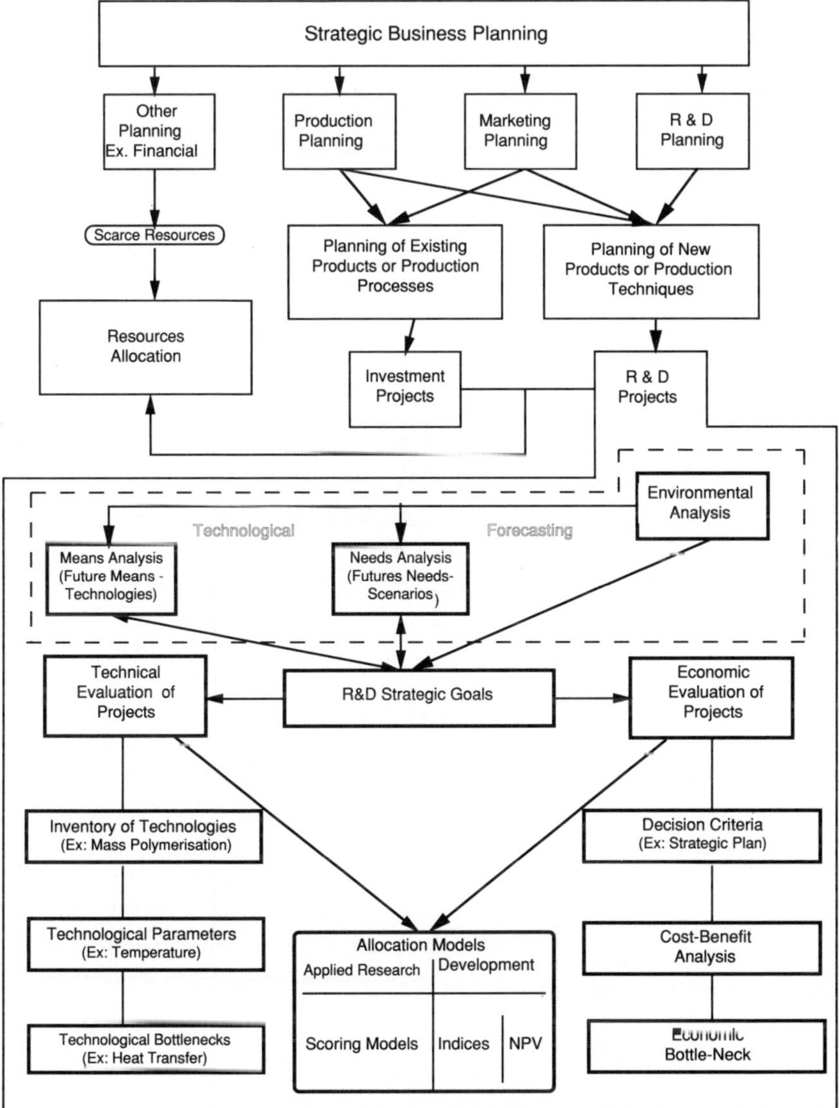

Figure 3.6   Bemelman's model of technology-strategy's genesis.

The matrixes described by Coombs and Richards work as follows. 'The Business Quality Matrix is used to differentiate into eight different business potential categories... Having differentiated all the business by this method, they are then mapped onto a standard 2x2 BCG matrix. The mapping is such that it recognises that the place on the grid should take into account the dynamic nature of business, and that businesses can be

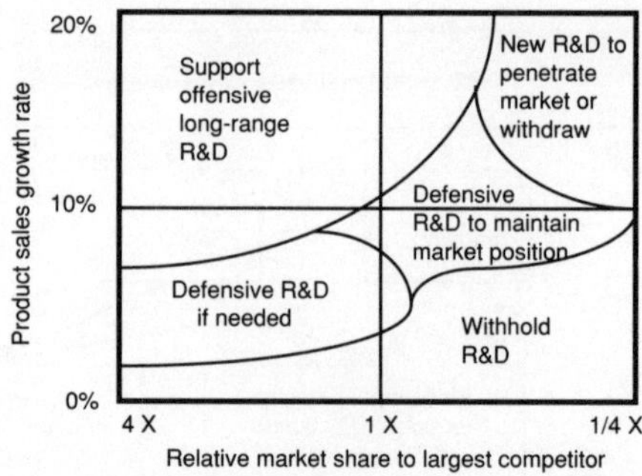

a) R&D as a function of business attractiveness

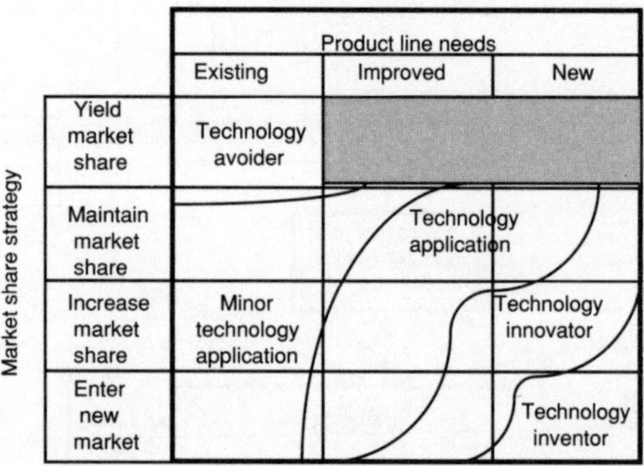

b) R&D posture in accordance with product lines and market share

Market share strategy	Product line needs		
	Existing	Improved	New
Yield market share	Reduce products		
Maintain market share	Consolidate products	Minor product improvement	New product existing specifications
Increase market share	Reduce cost of products	Product improvement	New product improved specification
Enter new market	Apply products to new use	Product redesign	Diversification
	Marketing: market strategy	Engineering: product development	Research: basic and applied R&D

c) Product line and technology development

Figure 3.7   Decision-aids for technology-strategy generation (adapted from Bitondo and Frohman, 1981).

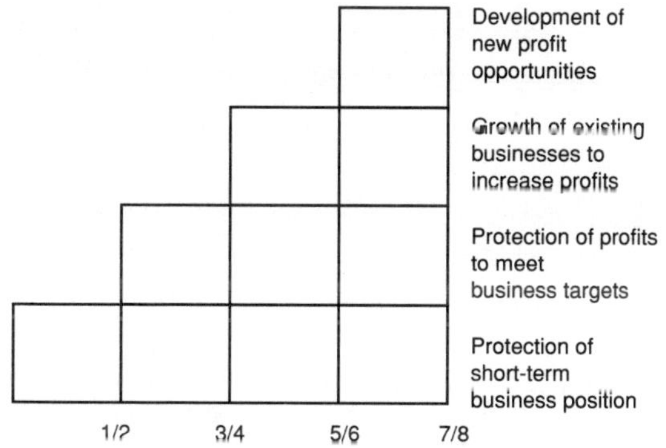

a) Business quality matrix

b) Business differentiation matrix

c) Business differentiation category and R&T effort

Figure 3.8   Decision-aids for technology-strategy generation (adapted from Coombs and Richards, 1990).

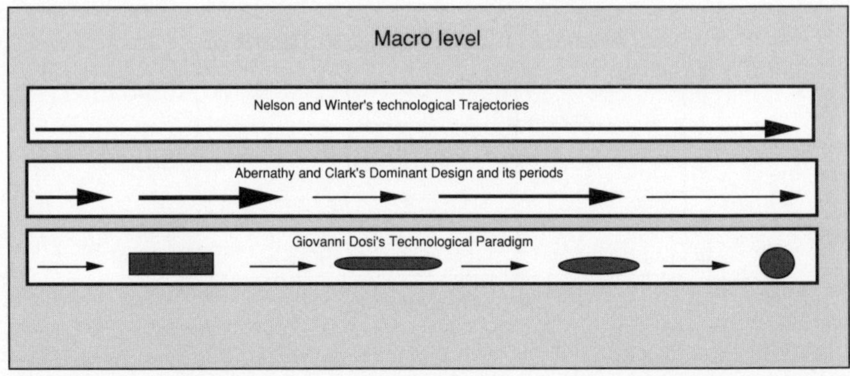

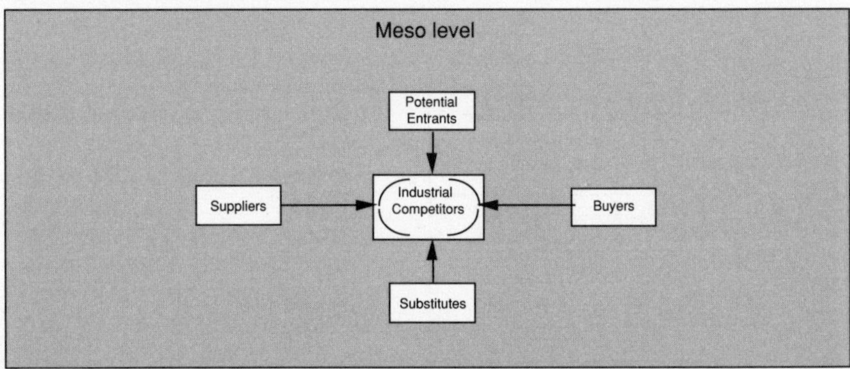

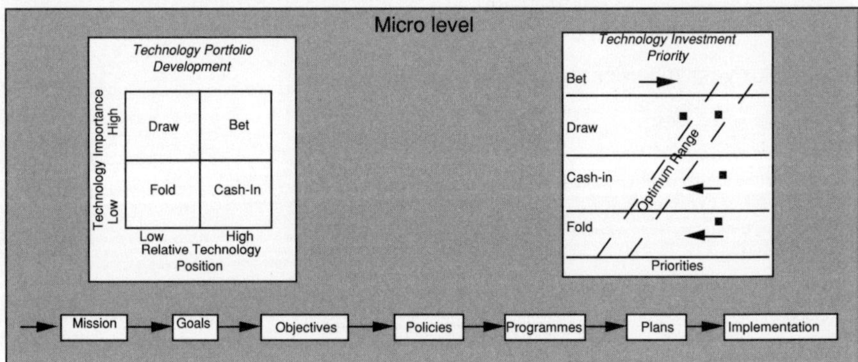

Figure 3.9 Mulder and Vergragt's model used as a framework for technology-strategy analysis.

located at any point in space within the grid.' Each business differentiation category requires a different level of total R & T (research and technology) effort, which is built up from consideration of the business requirement and the R & T category. The model assumes all businesses require the minimum level of R & T effort, but only the Strong Growth and New Business require the development of a new 'Profit Opportunities' R & T category. The level of each category is based upon a percentage of sales income and is generated from previous expenditure information. 'The R & T effort is a cumulative total of the categories allocated to each business.'

From the frequency with which these concepts, ideas and tools occur in the literature it is important to search for their utilization in the market-place (see Chapters 5 to 8).

The work of Coombs and Richards cited above is the culmination of comprehensive research on technology-strategy formulation. It clearly demonstrates the possibility of identifying technology strategy as something distinct from but related to business strategy. They make another important distinction between the strategic analysis of a product portfolio and the technology base in a business. For each of its case-studies the research was conducted using company documentation (over the previous five years), interviews at various levels and attendance as observers at management meetings and committees and even board meetings. Such a deep probing provides an excellent source of information on technology-strategy formulation (or micro-level) analysis.[7]

3.5 Conclusion

Figure 3.9 illustrates the use of Mulder and Vergragt's model as a framework to fit more specific models such as those of Nelson and Winter, Abernathy and Clark, Dosi, Porter and Wilkinson and several others reviewed throughout this chapter. Models such as those described by Wilkinson and Coombs have been adopted in some companies. It can be assumed that Porter's guidelines from general strategies are also in use. The main questions at this level are also related to the extent to which these techniques have diffused. Which gives better results? Are the use of these techniques associated with the cultural changes cited by Mulder and Vergragt? Are they in some way associated with a new paradigm? Are there patterns of approaches to these questions? Do these patterns change only with the size of the business or also with their position within the supply-chain? Can stereotypes be observed? What is the proportion of managers who see technologies and products from different perspectives? In those companies that actually have a formal technology strategy, what is the process of strategy generation? Are tools such as technological forecasting, technology auditory and portfolio analysis in use? How often? Are there any others? What are their results?

Notes

1. This second concept is derived from that of Marquis and Mayers (1969), who defined innovation as 'the unit of technical change'.
2. It is not suggested that the definition of 'internal innovation' is a refinement of Schumpeter's concept in any sense. It will be used only as a complementary idea. It is a definition restricted to the units of analysis of this work which are

firms and their individual SBUs and not the global economy in which Schumpeter focused his work.

3. Mulder and Vergragt also cite the work of Henk van den Belt and Hip (1987) as an important 'sociological' complement of Nelson and Winter's economic model of natural trajectories.

4. The cultural question seems to have an important role to perform in relation to technology. Cultural development within chemical companies has evolved through managerial innovations such as the concepts of 'just in time' and 'total quality'. These factors associated with the development and dissemination of digital technologies led to the so-called new 'techno-economic paradigm' as shown by Freeman (1989, p. 20):

'(1) Information-intensity rather than energy-intensity in products and process...

'(2) Flexibility in process investment and in product mix rather than the dedicated capital equipment and standardised products range...

'(3) "Systemation" not just "automation" ... The new paradigm uses information technology and communication technology to integrate the production plant...'

5. The technology matrices are usually used together with the classic business matrices (see Chapter I). While the latter are clearly situated at the meso-level of the Mulder and Vergragt model the same cannot be said about the technology matrices which are a tool developed to aid in the generation of technology and R & D strategies.

6. The concept of technological forecasting seems to be going out of the technology vocabulary 'hit-parade'. As described in the previous chapter in relation to 'strategic planning', the problem seems to be again a mismatch between a concept and its methodologies. This concept of technological forecasting is a very sound one (as good as any marketing forecasting for example). Nevertheless, none of the methodologies (see Twiss 1986) seems to have been broadly accepted as appropriate.

7. Coombs and Richards highlighted the potential mismatch that those methodologies (based upon matrices derived from business sources) can bring up in relation to technology and product life-cycles.

4 Chemical industry, innovation and strategy

4.1 Introduction

Innovation in the chemical industry has undergone impressive changes in its patterns, from the fundamental role played by entrepreneurs and bright 'back yard inventors' to the domain of the multinationals and their huge teams of professional researchers. This chapter describes, through a historical approach, the evolution of the industry and the role of innovation in its growth.

The first part of this history shows the industry's birth and the fast pace with which its intrinsic scientific character provided a leading edge to the companies (and countries) whose research was first professionalized.

With the industry's growth, its nature clearly became not merely science-intensive but also capital-intensive. In the chapter's second part, this capital-intensiveness is shown as a major factor in the emergence of the petro-chemical sector and its innovation pattern.

The third, and last, part shows the innovation process within modern firms. It is intended to shine some light on the present pattern, the industry's main challenges and the different kinds of innovations and innovators. The chapter concludes with some open queries about the near future, the possible emergence of a new techno-economic paradigm and its eventual consequences for industrial innovation.

4.2 The chemical industry evolution

As stated by Reuben and Burstall (1973) 'for many centuries men have wanted soaps, dyes, medicines, love-philtres and poisons'. Although it may not be true for soaps, at least medicines and dyes have been used not just for many centuries but for several thousands of years. The significance and scale, however, became typically industrial only after the beginning of the industrial revolution at the end of the 18th century.

Table 4.1   Chronology of some of the main inventions and innovations
in the chemical industry.

1616 – Window glass
1623 – Mining with gunpowder
1624 – Patent law protecting inventions in England
1688 – Distillation of gas from coal
1704 – Prussian blue
1746 – Sulphuric acid (lead chamber)
1750 – Alizarin
1779 – Glycerine discovered
1780 – Elementary quantitative analysis (C,H,O)
1784 – Leblanc-process invention
1785 – Chlorine as bleaching agent
1794 – Bleaching powder
1819 – Naphtalene found on coal
1824 – Methanol identification
1826 – Aniline synthesis
1833 – Electrolysis demonstrated by Faraday
1834 – Discovery of phenol
1835 – Vinyl chloride, first preparation
1836 – Acetylene discovered
1839 – Goodyear's vulcanization
1843 – Superphosphates
1845 – Nitroglycerine
1845 – Petroleum
1846 – Nitrocellulose
1848 – Acetic acid from wood
1850 – Acrylonitrile, first preparation
1855 – Oil-cracking demonstration at Yale
1856 – Mauveine
1857 – Methanol synthesis
1860 – Oil cracking (atmospheric)
1860 – Isoprene isolation
1863 – Alkyd resin invention
1865 – Benzene structure determined
1865 – Parkesine (celluloid precursor)
1865 – Solvay's process, first commercial plant
1867 – Nitroglycerinated explosives
1869 – Celluloid invention
1870 – Paper from wood pulp
1871 – Polyvynil chloride invention
1878 – Alkylation of aromatics
1882 – Viscose rayon
1884 – Urea-formaldehyde resin invention
1884 – Nitrocellulose rayon
1888 – Formaldehyde from methanol
1889 – Smokeless powder
1890 – United Alkali Company
1890 – H_2SO_4 contact process
1891 – Phenolic resin discovered
1891 – Phenol-formaldehyde reaction observed

1891 – Epoxy resin invention
1892 – Acetylene from carbide
1892 – Viscose invention
1893 – Cyclohexane synthesis
1897 – Isoprene first synthesis
1898 – Aspirin
1899 – Galalith production
1905 – Viscose industrial production
1905 – Acetate cellulose
1907 – Phenol commercial production
1908 – Haber-process invention
1909 – Burton-refining-process development
1909 – Phenolic resin production
1911 – Polyvynil acetate invention
1912 – Cellophane patented
1912 – Vinyl chloride production patent
1913 – Burton-refining-process operation
1913 – Haber plant start-up
1916 – Invention of isopropyl alcohol petrochemical process
1917 – Cellophane production
1919 – Petrochemical isopropyl-alcohol production
1919 – Brunner Mond inspects Haber process plant in Germany
1920 – Buble plate distillation column used in oil refining
1921 – Tariffs protection on chemicals' import (UK)
1922 – Tariffs protection on chemicals' import (USA)
1922 – Polyacrilonitrile
1923 – Synthetic methanol production
1923 – Dubs-refining-process production
1924 – British Haber plant operational
1926 – Alkyd resin commercial production
1926 – IG FARBEN's incorporation
1926 – ICI's incorporation
1927 – Vinyl chloride commercial production
1928 – Polyvynil acetate production
1928 – Urea-formaldehyde resin production
1930 – Polystyrene commercial production
1931 – Ethylene oxide from ethylene
1931 – Poly-(2-chlorobutadiene) production
1931 – UK tariff protection extended
1931 – Polyvynil chloride commercial production
1932 – Nylon invention
1933 – Polyethylene invention
1935 – Methylmethacrylate
1935 – Poly-(butadiene-styrene) commercial production
1935 – Melamine-formaldehyde resin production
1936 – Houdry catalytic-cracking production
1936 – Alkylation process invented
1937 – Poly-(isobutene-butadiene) production
1937 – Poly-(butadiene-acrylonitrile) production
1938 – Catalytic Research Association incorporated
1938 – Polyethylene pilot production
1938 – Acrylonitrile production from hydrocyanic acid

1938 – Nylon 6/6 production
1938 – Nylon 6 production
1940 – Polyester fibres invention
1940 – Nylon commercial production
1942 – Cyclohexane production
1942 – Fluid catalytic cracking
1944 – Acrylonitrile produced from acetylene
1944 – Isoprene production
1945 – IG FARBEN broken up
1948 – Acrylic fibres commercial production
1948 – Epoxy resin commercial production
1949 – Polyester fibres production
1950 – Teflon
1952 – UK changes from ethanol to oil-based ethylene
1953 – Polyamide fibres
1954 – Polypropylene
1955 – Polyisoprene
1956 – High-density polyethylene
1957 – Polypropylene commercial production
1958 – Polycarbonate
1960 – Acrylonitrile production from propylene
1960 – Use of zeolites as catalyst
1961 – Benzene from toluene
1962 – Naphtha reforming
1964 – Caprolactam
1966 – Low-pressure methanol
1967 – Propylene oxide
1968 – Butadiene extraction
1968 – Adiponitrile from electrodimerization
1970 – Hydrocracking using zeolites

4.2.1 First years

The technical basis for the flourishing of the chemical industry was provided by Leblanc in 1784–89. His process involved an initial reaction between sulphuric acid and sodium chloride producing sodium sulphate,
which was further reacted with coal and limestone to produce sodium
carbonate. Soda ash could be obtained from the reaction of this latter
product with slaked lime (calcium hydroxide).[1] Soda ash was used then
mainly for the production of soap. From 1800 to 1900 the consumption per
capita of soap quadrupled, providing the market for the expansion of the
'alkali trade'. Another strong demand was generated by the housing boom
from 1870 that brought an increased consumption of sodium carbonate
which was necessary for glass manufacture. Other important innovations
which combined with the industry consolidation were: bleaching powder in
1794 (fundamental to the growth of the cotton industry), the lead chamber
process for sulphuric-acid production, paper from wood pulp in 1870,
superphosphates in 1843 (the first fertilizer demand). The growing consumption of nitric acid for the production of explosives based upon nitrocellulose (invented in 1843) and nitroglycerin (dominated by Nobel and

66

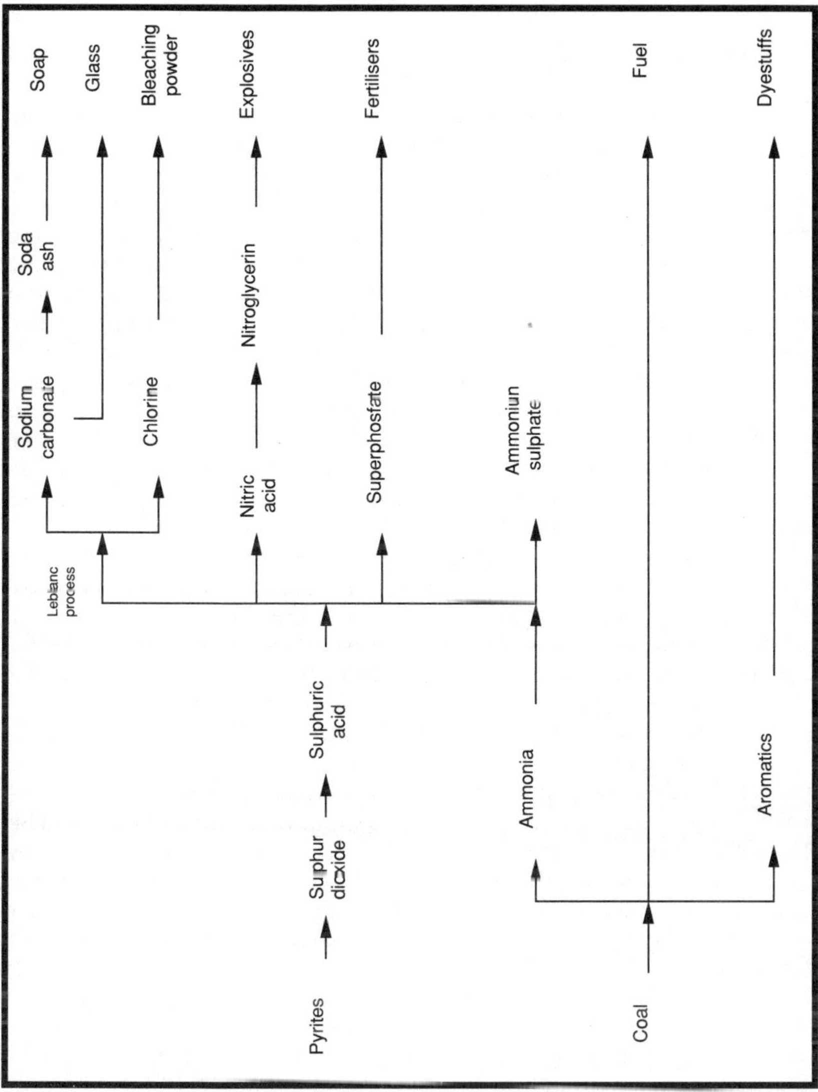

Figure 4.1 The chemical industry in the 19th century.

67

introduced into the market from 1867) also helped to elevate the consumption of sulphuric acid, providing some 'critical mass' to the new industry. Table 4.1 shows some of the most important inventions and innovations in the chemical industry. Figure 4.1 shows the structure of the chemical industry in the 19th century.

4.2.2 Dyestuffs and the German domain

Owing to the economical conditions generated by the industrial revolution, the British chemical industry started ahead of its main competitors. Such an advantage, however, became a disadvantage after a few decades. The British industry was heavily based on the first, and consequently rudimentary, processes such as that of Leblanc. British manufacturers using Leblanc's process were determined to compete but were blind to the evolution of the technology. As a counterpart to the start-up of the first Solvay plant in the United Kingdom in 1874, the 'Leblanc manufacturers' created the United Alkali Company in 1890 which stuck to the already obsolete technology.[2] The German industry capitalized upon their late start and introduced electrolytic processes rapidly. During this period, in accordance with Reuben and Burstall (1973), German universities were able to generate large numbers of research-trained chemists, as opposed to what happened in Britain. The dyestuff industry is clear evidence of this difference. While the first synthetic dyestuff was invented (and produced) in the United Kingdom in the 1850s, the Germans flooded the market with innovations such as alizarine in the 1860s and azo and sulphur dyes in the 1870s and 1880s. As described by Freeman (1982), the Germans also pioneered the creation of industrial research and development in the 1870s. In 1880, Germany already produced a third of the world's dyestuff output and, 20 years later, 80 per cent. During the same period the British educational system was unable to produce a significant number of technologists or research-trained chemists.[3]

By 1920 British dyestuff production represented less than 20 per cent of the world output and consisted of the cheapest and simplest processes. The high scientific level of German chemical companies was accompanied by that of the Swiss industry, leading to the successful development of synthetic medicines. According to Freeman, at the end of the 19th century Switzerland was already exporting 93 per cent of its drugs and dyestuff production.

The most astonishing German innovation in that period, however, was still to come. The Haber (or Haber–Bosh) process for direct reaction of nitrogen and hydrogen at a high temperature and pressure (600° centigrade and 200 atms) over a catalyst to produce ammonia was the first 'modern' chemical process. The importance of this process can be illustrated by the technical lead that it provided in terms of high-pressure technology. While the first Haber-process plant was operating in Germany at over 200 atm in 1913, the Americans were having serious engineering problems with their contemporary Burton oil-refining process which operated at 5 atms. Even by the 1930s the Americans were still having to import valves and compressors from Germany to run their catalytic cracking plants at less than 30 atm. The Haber process also had an essential role to play during the First World War. Had Germany not had this synthetic source of fixed nitrogen, they would probably have run out of food and explosives by 1916 (see Reuben and Burstall, 1973). The war provided the economic conditions for the development of several other innovations, which will be discussed further in Section 4.2.4.

4.2.3 The oil industry and American growth

At the end of the 19th century the American chemical industry was not yet globally important, but, owing to fast economic growth, by the 1920s it had possibly already become the second largest in the world. However, in spite of some important inventions and innovations such as rubber vulcanization, the Americans were way behind the Germans in terms of technology.

America's main contributions started with growth of the oil industry, which was its main source of organic raw materials (in contrast to the importance of coal in Europe). To be useful, petroleum must be refined as it is a complex mixture of several different substances. The earliest refineries used to split it into just four components: impure gasoline, kerosene, lubricating oil and fuel oil. Once it became clear that the demand for the different parts was not in the same proportion as they occur, chemists were soon working on how to link supply and demand. The solution to such a linkage was reached in 1855 (see Table 4.1) when the ability to break down petroleum's large molecules into smaller ones was first demonstrated. According to Enos (1962a) this principle was used to yield kerosene from 1860 (at atmospheric pressure). At the beginning of the 20th century the demand for gasoline was growing owing to the advent of the automobile industry, while the substitution of kerosene lamps by electrical bulbs reduced the demand for that fraction. The technical basis for matching these demands was first provided by a patent obtained by two English chemists. The patent described the process by which oil, through submission to elevated temperature and pressure, boils in gasoline rather than in the kerosene range. The first industrial pressurized cracking process was developed by William Burton in 1909. Burton was a refinery manager who had his own chemistry laboratory as a boy. He not only sorted out the engineering teething troubles but also improved the process by cracking just the gas-oil portion.[4] The first commercial production started in 1913 and resulted in a double yield of gasoline and provided a total income between profits and royalties of US$ 150 million (see Enos, 1962a and b). Among several other innovations in the cracking process, the 'Tube and Tank' developed by Standard Oil of New Jersey (now Exxon), the Dubbs and the Cross processes were the more important (all continuous thermal-cracking processes).[5]

During the First World War, several major innovations were generated in the chemical industry, the main one being the birth of the petrochemical industry, as discussed below.

4.2.4 The First World War and the petrochemical industry's birth

The cutoff in the supply of German organic chemicals led to a strong growth of the British and American chemical industries. In Germany the strong demand for explosives switched the production capacity from dyestuffs with remarkable speed. It consolidated the Haber process and the synthesis of nitric acid from ammonia (produced through the Haber process). Also during this period the Germans produced the first synthetic rubber (poly-2,3-dimethyl rubber) and some adhesives based upon polyvinylacetate.

To exemplify the revival of the British chemical industry during the First World War, it can be said that its production of dyestuff grew by 400 per cent from 1913 to 1919. The Weizmann process for the production of butanol through the fermentation of corn was invented in Britain (although it had a major importance only in the American war effort). Butanol was an important intermediate in the production of acetone that was used in the gelatinization of explosives and powders.

Acetone was also the motivation behind the first petrochemical-process invention. In 1917 an American chemist named Carlton Ellis invented a process for the manufacture of isopropyl alcohol (a raw material for acetone production) from oil refinery waste gases. The process's main steps were the absorption of propylene in sulphuric acid, followed by hydrolysis and distillation. The war finished before the industrial use of the process was acquired by the Standard Oil of New Jersey to be used as a gasoline 'extender'. The first plant was commissioned in 1919 as the world's first large petrochemical plant.

In spite of the pioneering role of the Ellis process, most of the technology which caused petrochemicals to flourish was generated by the work of Dr George Curme (see Miller, 1969) at the Mellon Institute in Pittsburg. One of the main factors behind that work was the war. The ethylene generated in a small-scale unit (whose main product was acetylene) was an interesting raw material for the production of ethylene chlorhydrin, which, in his terms, could be transformed into mustard gas. By the end of the war in 1919, Carbide (which was supporting Curme's work) had begun several projects to transform ethylene chlorhydrin into commercially useful products.

Also in 1919, the Haber plants in Germany were carefully inspected by allied producers. The German technological lead was so great that it took seven years for the British leading producer Brunner and Mond to be able to imitate the process after the inspection of the BASF unit in Opau. Similar times were also taken by American and French producers.

In 1920, the last step towards the technological sustaining of the petrochemical industry was given at the Linde Air Products Company where ethylene was obtained (at laboratory scale) by the cracking of ethane/propane mixtures. The product was transported to the Mellon Institute where it was transformed into ethylene dichloride, ethylene chloridrin and ethylene glycol.

The performance and importance of the chemical industry left deep

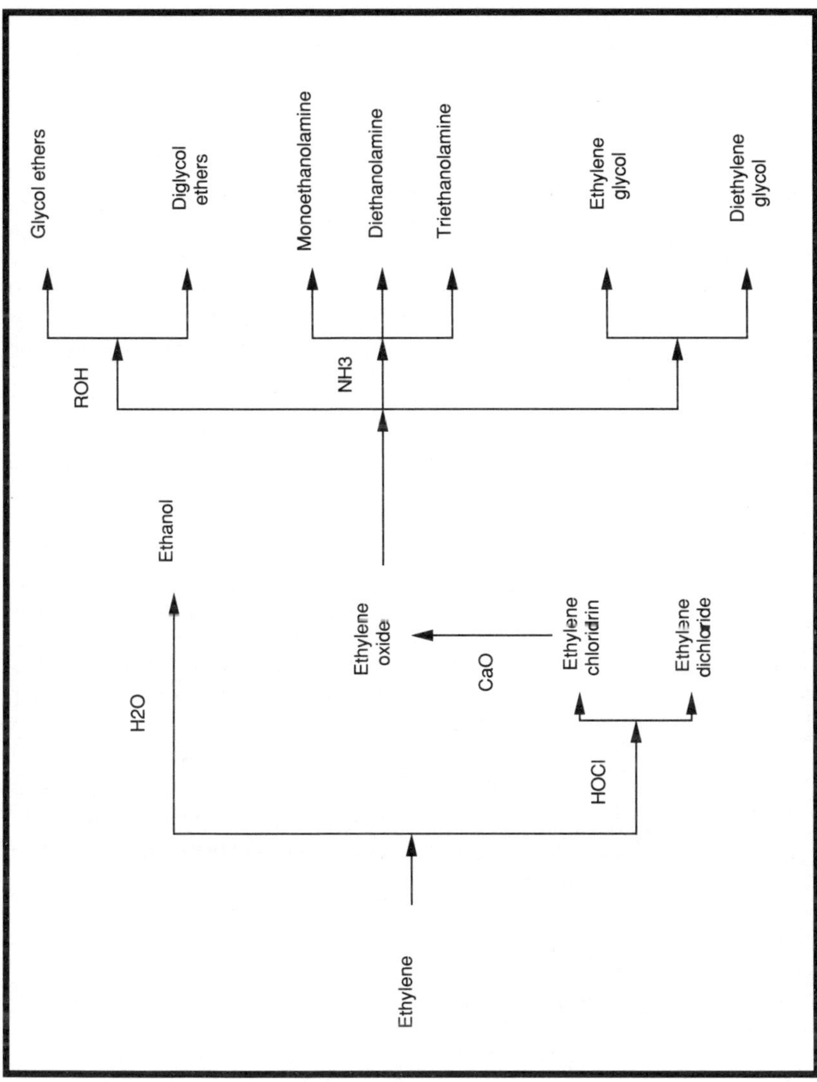

Figure 4.2 Earlier ethylene derivatives.

impressions on the British and American governments, leading to the creation of import duties in both countries in 1921 and 1922, respectively. America's legislation, in particular, virtually excluded all the imports for the inter-war era. This process resulted in the creation of ICI in England.[6] The same process resulted in the amalgamation of BASF, Hoechst and Bayer (among others) in Germany to form the most innovative chemical company in the world, IG Farben.

4.2.5 The inter-war period

The first years after the war were marked by depression and, in comparison with the war years, lacked innovations. One of the most important exceptions was the BASF process for methanol production which started industrial operation in 1923. Also in 1923, based upon the technology developed at the Mellon Institute, Carbide built the world's first major petrochemical complex. By 1925 the commercial production of a series of ethylene derivatives was established, as shown in Figure 4.2.

From 1926 a new wave of innovations was generated. In terms of thermosetting resins General Electric introduced the alkidic resins (1926), followed by the introduction of urea-formaldehyde by British Cyanide in 1928. Also in 1928, Carbide in the USA and IG Farben in Germany started commercial production of polyvinyl acetate.

As it was clear that dyestuffs had become a mature industry, the board of IG Farben decided to re-direct its efforts to the research of big molecules, the polymers. From 1925 to 1939 IG Farben's investment in research and development averaged more than 7 per cent of its turnover, even excluding donations to universities and capital investment in new laboratories as described by Freeman (1982). As a consequence, IG accounted for over a third of all patents related to polymers taken by the 30 largest companies and none of them, with the exception of Du Pont, had more than a sixth of that number of patents. Notwithstanding this indicator of inventiveness IG was also an impressive innovator pioneering the commercial operation of several new processes and products which are shown below:

Vinyl chloride monomer – 1927
Polyvinyl acetate – 1928
Polystyrene – 1930
Polyvinyl chloride – 1931
Melamine-formaldehyde resin – 1935
Poly-(butadiene-styrene) – 1935
Poly-(butadiene-acrylonitrile) – 1937
Acrylonitrile – 1938
Nylon 6 – 1939

During this fertile period of innovation great contributions were also made by many other producers. Among those innovative firms, Du Pont and ICI were of major importance. ICI was responsible for the discovery of methylmethacrylate in 1935, polyethylene in 1933 and its first industrial production in 1938. Du Pont, from the basic research conducted by Dr Carothers, invented nylon 6/6 in 1938 and started its production in 1940.

In terms of basic and intermediate petrochemicals, significant progress was made by Carbide and Carbon Chemicals Corporation (now Union Carbide). According to Goldstein and Waddams (1967), Carbide produced five compounds derived from ethylene in 1926, increasing this number to 41 by 1939. These data show the evolution of ethylene to the category of first and main (up to now) 'building block' of the petrochemical industry.

The 1930s also marked a second generation of major innovations in the oil industry. While the first wave of innovations (Dubbs, Tube and Tank and Cross) were all concerned with the improvement of the continuous cracking process and to increase the yield of motor gasoline, the second was related more to the quality of gasoline through the use of catalytic cracking.

The first successful catalytic process was developed by the French inventor Eugene Houdry, who started working on the subject just after the First World War. Unlike the other inventors cited above, Houdry was not an employee of any company. He spent at least US$3 million of his own fortune to develop the process (Freeman, 1982). During the last years of development two oil companies (Sun Oil and Socony) associated themselves with Houdry, providing technical and financial support. The Houdry original design was a semi-continuous process since each of the main vessels operated in successive cycles of cracking and catalyst regeneration. The first plant was started 1936.

Threatened by the advantages of the Houdry process, a group of companies created Catalytic Research Associates to develop a continuous process of catalytic cracking. The group was originally formed by Kellog, IG Farben, Standard Oil of Indiana and Standard Oil of New Jersey (Exxon). Soon the group was reinforced by Shell, Anglo–Iranian (BP), Texaco and UOP. In spite of the IG Farben withdrawal the research involved some 1,000 researchers and, at a cost of US$30 million, it was successful in 1942. In contrast to the Houdry process the new one was continuous, using a fluid bed catalyst.[7]

4.3 The petrochemical industry evolution

4.3.1 The Second World War

The outbreak of the war deeply affected the structure of the chemical industry. One of the best examples was the growth in synthetic-rubber production. Germany had an installed capacity for 5,000 tonnes of Buna-S in 1937, but at the end of the war that capacity had already reached some 150,000 tonnes per year. In the United States, on the other hand, the war effort led to the creation of another enormous cooperative research and development project. With the participation of Standard Oil of New Jersey, Dow Chemical, Goodyear, Goodrich, Firestone, Monsanto, etc. the American production of synthetic rubber jumped from a negligible amount in

Table 4.2 Alternative routes for chemical products
(adapted from Goldstein and Waddams, 1967).

Chemical	Petroleum source	Alternative sources
Methane	Refinery light gases (natural gas)	Coal, as by-product of separation of coke oven gases or of coal hydrogenation
Ammonia	Methane Light liquid hydrocarbons	From coal via water-gas
Ethylene	Pyrolyses of gaseous or liquid hydrocarbons	Dehydration of ethyl alcohol. By-product in fractional distillation of coke oven gas
Ethylene glycol	Ethylene	From ethylene made as above. From coal via carbon monoxide and formaldehyde
Acetylene	Methane Light liquid hydrocarbons	Calcium carbide. Methane from coal by partial combustion and by arc process
Ethyl alcohol	Ethylene	Fermentation of molasses
Acetaldehyde	Synthetic ethyl alcohol. Co-product of paraffin gas oxidation. Direct oxidation of ethylene	Fermentation ethyl alcohol, or acetylene from carbide
Acetone	Propylene	Wood distillation. Pyrolyses of acetic acid or by acetylene-steam reaction
Glycerol	Propylene	By-product of soap manufacture
Butadiene	Butane. Synthetic ethyl alcohol. By-product of ethylene by pyrolyses of liquid hydrocarbons	Ethyl alcohol. Acetaldehyde via 1:3-butanediol. Acetylene and formaldehyde from coal via 1:4-butanediol. From 2:3-butanediol by fermentation
Aromatic hydrocarbons	Aromatic-rich and naphthenic-rich fractions by catalytic reforming and direct extraction . Hydrodealkilation	By-products of coal tar distillation

1941 to 760,000 tonnes in 1945.

Other important events in the industry during that period were:

– Industrial production of acetone and derivatives (1936, Shell);
– Manufacture of vinyl acetate (and chloride) (1936, Carbide);
– Pilot production of polyethylene (1937, ICI);
– Full-scale production of styrene and butadiene (1938–9, Dow);
– Production of aromatics from petroleum (1940, Shell and Humble);
– Commercial production of nylon (1940, Du Pont);
– Invention of polyester fibres (1940, Calico Printers);
– Cyclohexane production (1942, Du Pont);   ·
– Acrylonitrile produced from acetylene (1944, IG Farben);
– Manufacturing of isoprene (1944, Standard Oil of New Jersey);

The consumption of ethylene during the period jumped from 63,000 tonnes per year to 135,000 (Miller, 1969, p. 35) for production of ethanol; 500 tonnes per year to 58,000 for ethylbenzene production; 41,500 tonnes per year to 78,000 for ethylene oxide; 9,000 tonnes per year to 27,000 for ethylene dichloride and 3,000 tonnes per year to 28,500 for ethyl chloride. In Germany at the same time the ethylene production (although essentially not petrochemical) reached 108,000 tonnes in 1943.

While the first war led to the creation of the petrochemical industry, the second led to its widespread expansion. During the period process engineering was significantly enhanced. Separation processes such as extractive distillation, liquid-liquid extraction and continuous absorption were improved. New materials led oil cracking to more severe conditions, increasing the proportion of olefine by-products. During the period petroleum also became a source of aromatics, which had been the last coal economic advantage in major chemical raw materials. At the same time paraffins, acetylene and di-olefins were brought to the economic scene from oil.

4.3.2 The post-war period

Up to the end of the war, the petrochemical industry was significant only in the United States. In spite of the fact that several technological developments in the industry had taken place in Germany (direct oxidation of olefins, reaction of olefins with carbon monoxide to alcohol production, high-pressure technologies, etc.), only the United States had a structure of demand which enabled the use of petroleum as an important raw material.[8] The European industry was still heavily based upon coal. During the 1950s all major European economies shifted (at least partially) their source of chemical raw materials towards petroleum. Table 4.2 (adapted from Goldstein and Waddams) shows some of the products which had their source of raw material changed.

From the end of the war to the beginning of the 1950s, in spite of the economic problems, the chemical industry kept innovating. In 1948 Du Pont launched the production of acrylic fibres and of epoxy-resins. In 1949, ICI started its polyester fibres production.

Table 4.3 Per capita consumption of plastics in industrial nations (adapted from Reuben and Burstall, 1973).

	1950	1955	1960	1961	1962	1963	1964	1965	1966	1967	1968	1969
USA	6.4	8.0	10.7	17.0	18.1	19.1	20.8	24.6	28.5	28.7	32.5	34.7
UK	2.5	5.2	9.1	9.1	10.4	11.5	13.9	15.1	15.5	17.3	20.4	21.7
W. Germany	1.9	5.8	15.0	15.0	21.4	19.5	24.5	26.7	28.5	31.6	40.4	49.9
France	0.9	2.8	7.5	7.5	9.5	10.8	12.5	14.1	17.2	18.9	21.6	26.8
Italy	0.6	1.7	5.0	5.0	8.6	10.2	11.0	11.8	14.7	17.7	19.3	23.5
Japan	0.2	1.4	3.8	3.8	10.6	10.8	13.6	14.2	17.0	23.4	29.0	32.8

(kg per head)

* Including cellulosics but excluding elastomers and fibres.

From the 1950s the petrochemical industry was no longer exclusively American, although in 1956 American production still represented some 87 per cent of the world capacity (Guglielmo, 1962).

During the following period (from 1954) Germany, the United Kingdom, France, the Netherlands, Italy and Japan became important players in the petrochemical sector. The period marked the building up of important refineries in Europe. These refineries were built to yield as much middle and heavy fraction (for power generation) as possible. With the relatively small demand for gasoline, naphtha (oil fraction with boiling point between 40° and 220° centigrade) became a cheap by-product. In spite of their initial small size, the European crackers were able to produce a cheap ethylene whose price dropped two-and-a-half fold in real terms (Arni, 1982). This combination of price reduction with availability led to strong competition with acetylene, which until that time was the chemical industry's main intermediate feedstock (derived from coal).

The change of the production pattern in the United Kingdom may be taken as typical: while British chemical production was based 60 per cent on coal and less then 9 per cent on petroleum in 1949, by 1963 it was already based 60 per cent on petroleum (Reuben and Burnstall, 1973, p. 33).

The second important change during that period was the beginning of the industry's internationalization as Japan's refining and petrochemical industries closely followed the American and (mainly) European stages.

Since the beginning of the 1950s most of the innovations were on the process side rather than new high-tonnage products. Polypropylene is one of the most illustrious exceptions, although even its discovery was based upon a process development.[9]

During this period the engineering firms such us UOP, Kellog and Lummus, often based upon experience acquired in oil refining, were able to give some extra technological thrust to the industry. The same happened in Germany as engineering firms such as Uhde and Lurgi had a similar role which continues up to the present day. The industry's main thrust, however, was no longer in the scientific–technological field but rather in marketing. The commercial effort developed by the American corporations for the creation of consumer markets for polyethylene and polyvinylchloride opened the way to the huge markets for the different synthetic polymers still to come into commercialization.[10]

4.3.3 The golden era

As had happened to sulphuric acid in the past, ethylene became so important (as the main petrochemical feedstock) that its production started to be used as an economic growth indicator. 'The stage of development of a nation could be measured by its ethylene production capacity.' European ethylene production from 1965 to 1973 grew at an average of 21 per cent per year, while, during the same period, the Japanese capacity grew at 23.3 per cent (Arni, 1982). (Since the beginning of the 1960s synthetic polymers' production also grew at an explosive rate.) The kind of innovations, however, changed sharply from a balanced succession of products and process innovations to a large domain of process innovation.

From 1949 to 1969 the world synthetic polymers' production grew by an astonishing factor of 30. Table 4.3 (adapted from Reuben and Burstall, 1973, p. 35) shows the evolution of the per capita consumption of plastics in seven industrial nations. The growth of polymer's market was the main reason behind the petrochemical industry growth as a whole. Stobaugh (1988) estimated that the American petrochemical industry grew at a yearly average of 10 per cent from 1961 and 1973.

Synthetic fibres were one of the main applications for polymers and grew at equally astonishing rates, with each producer manufacturing several different types of fibres and selling each of them for any type of end use. In Chapter 6 a more specific and comprehensive review of the synthetic fibres' evolution is presented.

The golden era marked a rush of new entrants in the chemical industry in all industrialized countries, accompanied by lower prices,[11] although generating large profits and new applications. One of the main characteristics of that period was the exploitation of the economies of scale, which led to the construction of huge plants and to the realization of the industry's global nature. As a consequence, not just the industry but each major producer became international.

At the beginning of the 1970s, however, the industry's growth was already much smaller than in previous decades and some sectors, synthetic fibres for example, were already showing signs of maturity. At that time a common strategy to avoid the competition of new entrants was to build plants with a capacity considerably larger than the forecast demand, creating a calculated idle capacity in order to de-motivate possible new competitors. Among the companies using that strategy were the oil companies. Attracted by the high profits in the sector, the oil giants had integrated their process downstream and became competitors with the 'truly chemical companies'.

4.3.4 The oil shocks

The 1973 cutback of oil production by OPEC led to a sharp rise in prices in petrochemical raw materials and energy. Such factors, accompanied by the world economy recession, left the sector in its worst-ever situation. Installed capacity was much bigger than consumption, generating low prices allied to high costs. In several cases business revenues couldn't cover the variable costs, leading to several plants being closed. From 1973 to 1975 American petrochemical output fell by some 20 per cent. European and Japanese producers were even worse affected.

From 1976 to 1979, however, the world economy had a strong upturn and the 'readapted' petrochemical industry went back to investments based upon its historical growth indexes. At that time the industry structures of cost and of capital had already been affected by the first shock.

During the two previous years those companies with captive sources of raw material capitalized upon the oil embargo and the high prices of petrochemicals. Those companies were mainly the 'chemical arms' of oil companies. They invested heavily and made the industry more vertically integrated. In terms of cost structure the industry was deeply affected. While the fixed costs historically represented 70 to 80 per cent of the total, the post-shock structure was almost inverted, making the variable costs achieve and sometimes surpass 80 per cent.

The second oil shock in 1979 found the industry again with an enormous idle capacity. The industry's reaction had two main components. The first was the collusion in the formation of cartels, the second real restructuring. Among the main losers were the oil companies' chemical subsidiaries. By 1982 they were still sharply reducing their petrochemical operations. During the first half of that year Occidental Petroleum mothballed 20 per cent of its PVC capacity, Royal Dutch Shell cut 21 per cent of European ethylene and 15 per cent of its propylene capacities (*The Economist*, 1982). Exxon's chemical division shut 25 per cent, 25 per cent and 10 per cent, respectively, of its ethylene, polypropylene and low-density polyethylene capacities. Similar actions were taken by BP Chemicals, Gulf Oil and others. Unlike the first shock, however, the readjustment now had not just a quantitative but mainly a qualitative nature. Several firms, mainly in Japan and Europe, swapped businesses in accordance with their strengths and weaknesses. BP Chemicals, for example, swapped its PVC business (in which it was small) with ICI, receiving as counterpart its low-density polyethylene assets. Both companies strengthened their portfolios, shutting down in a much more modest and rational way than could be expected by the independent cut-downs. This type of 'qualitative' reduction of capacity was followed by several companies, particularly in the synthetic fibres sector (as described in Chapter 6).

The previous action, based upon the formation of cartels (*European Chemical News*, 1991a), failed owing to the ultimate changes in the growth pattern of the industry. As cited by Cook and Sharp (1991), the 'petrochemicals income multiplier (the percentage growth for every one per cent GDP) dropped from 3 to 1,4'.

4.4 The 1980s and 1990s

While the previous periods showed a clear trend of forward process integration by the oil companies, the same could be observed about the 'truly chemical firms'. As the former diversified from their condition of simple refiners and crackers to bulk plastic and chemical manufacturers, the latter moved further ahead to specialities, fine chemicals and pharmaceuticals During the same period competition in the petrochemicals markets was becoming stiffer following the emergence of new entrants, mainly less-developed and gas-rich countries. By 1984, however, the industry was again in the upward part of its already well-known cycle. With the economic boom of the two following years, demand grew very fast, accompanied by margins and profits, mainly benefiting the companies which remained heavily based on sound commodities. Such a situation didn't last long, and the commodities down-turn reappeared in 1988–89, getting really difficult by 1990–2. At that time the companies with relatively small problems were those with more forward integration. Once again the European industry was suffering from the over-expansion begun during the previous up-turn.

By 1990 ethylene consumption in Europe was about 14.7 million tonnes while capacity was around 16.3 million tonnes and increasing to a projected capacity of 19.5 million by 1993 (*Financial Times*, 1991a). At the same time the world market was being flooded by new capacity throughout the world (particularly Saudi Arabia and South Korea). At that time the only significant growth in consumption of basic petrochemicals was in the Far East, in countries such as South Korea, Taiwan, Hong Kong, Malaysia and latterly the south of mainland China.

While several of the most important innovations of this century were made in the chemical industry, they seem to be diminishing. There are several definitions and techniques for the measurement of industrial innovation which, in spite of differences, are all unanimous in showing the sharp decline in the major innovations in the chemical industry. For example, Bennet and Kline (1987) stated:

> From 1930 to early 1980's there were 63 major innovations in chemical products. In addition to developing major plastics and fibres, scientists invented fibreglass, silicones, herbicides, epoxies, flame retardants, high-nitrogen fertilizers and automotive converters. Of these, 40 were introduced in the 1930's and 1940's, 20 more in the 1950's and 1960's, and only three in the 1970's and 1980's to date.

Such statements can lead to an erroneous conclusion that chemical companies are not researching, or researching on a much smaller scale, which is far from the reality. Hoechst, Bayer, BASF and ICI, to cite but a few, expended in 1987 a total of £2.5 billion in R & D altogether, representing 6 per cent, 6 per cent, 4 per cent and 4 per cent, respectively, of their turnover. Since then those numbers are growing (in 1991 ICI, for example, invested 5.3 per cent of its turnover in R & D). What really happened is that the industry became technologically mature. The decades of 1920–40 marked the development of a new science, the polymer science. The more or less contemporaneous domain of catalysis (where there is still a lot to be developed) science added an odd thrust for innovation which cannot be expected to be continued indefinitely on the same scale. The industry's capability to 'engineer' molecules, however, is being applied to the so-called 'growth areas' of pharmaceuticals and agrochemicals.[12] Hoechst, Bayer and ICI invested more than 50 per cent of their R & D budget in those areas in a clear shift from their previous emphasis on basic chemicals.

Although it may appear to be the case, the petrochemical industry is not divorced from its science-based origins. Important innovations have been produced. Such innovations are, however, almost completely in the field of process innovation and rarely in product innovation. In spite of the high level of its capital intensiveness and of the industry's low margins, important innovations are on-stream. *The European Chemical News* (1991b), for example, cites new processes for polyethylene manufacturing such as Montedison's 'spherylene', BP's 'fluid bed' and Mitsui's 'super'. All those innovations are concerned with polyethylene, which is quite a mature product.[13]

The industry's (as a whole) heavier and most promising research investments, however, are concerned with an alternative feedstock to naphtha. Arni made a comprehensive review of the main alternatives country by country, product by product. Among the main alternatives are natural gas (mainly through methanol and ethylene), coal (through syngas and liquefaction), tar-sands and oil-shale. Although the steadily lower prices of crude oil changed the perspective considerably, natural gas is still under heavy investigation.[14] *The European Chemical News* (1991b) highlights the following processes:

- Oxidative coupling of methane (Arco, Phillips Petroleum, Union Carbide, IFP);
- Oligomerization to gasoline and diesel (Mobil);
- Direct oxidation to methanol/formaldehyde cracking to acetylene/ethylene (IFP);
- Chlorination/cracking to acetylene/ethylene (BP).

These research efforts are some of the most globally important, although the research is spread among big companies and through different types of products as described below.

4.4.1 The innovation process in the chemical industry

Industrial innovations can generally be divided into two main groups: the product innovations and the process innovations. While any product can, in principle, be produced by different processes the reciprocal is not necessarily true. This statement is particularly valid in the chemical industry where a product can be produced from different raw materials, processes, operations and conditions and still be exactly the same. There are, however, differences between the different segments of the industry in relation to their propensity to innovate in one or other mode. Oversimplifying, it can be said that commodity chemicals can be innovated mainly in terms of process, while specialities are equally 'suitable' for product and process innovations.[15]

Both types of innovations can still be subdivided into minor or major innovations. Particularly useful is Stobaugh's (1988) definition for process innovations: 'Major innovations are fundamentally different from existing processes, involving different raw-materials or radically different reaction conditions, such as the use of vapour-phase rather than liquid-phase reactions.'

Analysing nine petrochemical commodities Stobaugh found 180 innovations from their first commercial production up to 1974. Among those innovations only 29 were considered to be major ones, averaging three per product against 17 minor innovations. He also found that the number of major innovations declined steadily from the first to the fifth decade of the product's life. The number of minor innovations grew steadily up to the third decade, then declined for the other two decades.

Innovations in specialities are considerably more common since such products are usually sold on the basis of performance rather than composition. In so doing it is relatively easy to modify slightly their composition in order, for example, to attend to a specific requirement from a consumer. In such a context, a major product innovation could be, for example, the development of a new drug to be used in a herbicide, while the different formulations to be commercialized using that new technical product could be classified as minor product innovations.

Describing the innovation process in the chemical industry is a hard task; the number of variables and the degree of uncertainty of each one makes any model very inaccurate. In spite of this intrinsic inaccuracy, several steps are common to most of the developments. Some of these steps are illustrated below in order to provide a better understanding of the examples which will follow:

- The original idea conception;
- Literature review (which can be extended throughout the whole development);
- Development of analytical methods for monitoring and control of the experiments (including the synthesis of analytical standards);
- Laboratory experiments;
- Scale-up of the experiments and their engineering (depending on the process complexity, the scale-up can go straight from laboratory to industrial scale or pass through 'bench' and 'pilot' scale trials);
- Market evaluations and pilot production trials by customers;
- Industrial scale plant engineering and construction.

Throughout the whole development, economic evaluations are carried out as new data are produced at each stage. The number of looping and redefinitions of goals, capacities, locations, etc. can vary quite a lot. A typical example can be taken from Stobaugh (1988, pp. 29–37) where he describes 'Alpha's' project for a polymer development. The project was analysed up to the ninth year and was developed by a petrochemical company at a cost of $2 million.

> The project was initiated targeting the market for 'baked-on coatings'. At the end of year 9, 'non-baked coatings', 'hard' and 'soft elastomers', 'rigid' and 'flexible foams' and 'adhesives' markets had been evaluated. Twenty different processes had been evaluated, involving nine different monomers and dozens of raw materials. Estimates of possible selling prices went from 0.31 to 1.11 dollars per pound. Plant capacities were estimated from 500,000 to 50 million pounds per year.

At the start of this research project it was intended to make a decision about its commercialization within four years, although by the end of the ninth year it was still being rescheduled to several years later.

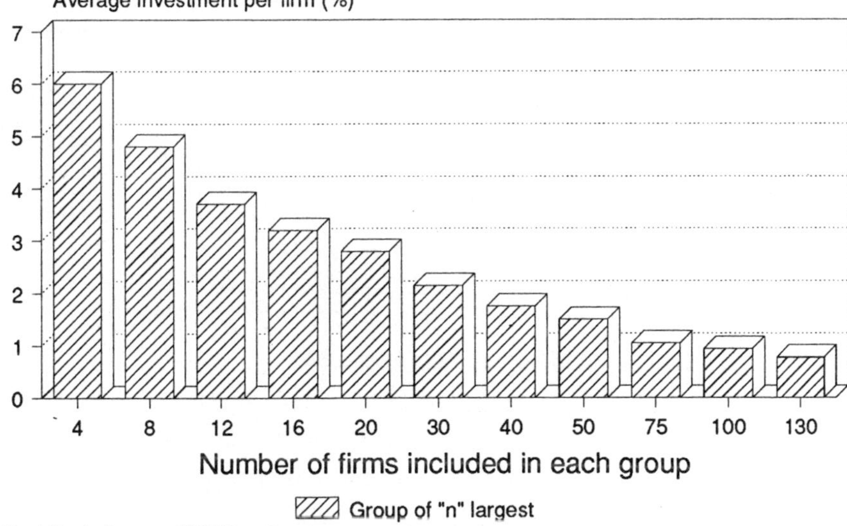

Figure 4.3 The concentration of industrial research and development investment in the USA (adapted from Soete, 1979).

The introduction of new processes and products is a risky initiative even for large and experienced innovators. As described by Freeman (1982), Du Pont had a major commercial failure in its synthetic leather (Corfam) project. Du Pont initiated its studies in poromeric films in the late 1930s. By 1956 field trials for applications in shoes were started, and after making more than 16,000 pairs of shoes in 200 different models as a part of the market research, they started industrial production in 1966.

In spite of the company's experience and use of the most modern technologies, management and marketing techniques (including computerized market simulations), the initiative was a complete failure. Although several million pairs of shoes have been sold, the project caused losses of US$100 million.

The examples above can be frightening to any investor, but the difficulties when overcome seem small in comparison with the benefits. Product innovation seems to be particularly profitable. As shown by Stobaugh, the average time span between a product's first commercialization and the arrival of a new entrant (of the nine commodities he analysed) was 5.7 years. Such a long monopoly (followed by an even longer oligopoly) can generate a more-than-profitable premium return. He estimated Ciba-Geigy's profits from the atrazin innovation, for example, as high as US$500 million in the United States alone. For Du Pont's nylon innovation, he estimated more than US$1 billion. Such high profitability was also observed for some major process innovations such as the catalytic ammoxidation of propylene for acrylonitrile production, for which Stobaugh related profits superior to US$700 million to the innovating company (see Chapter 5).

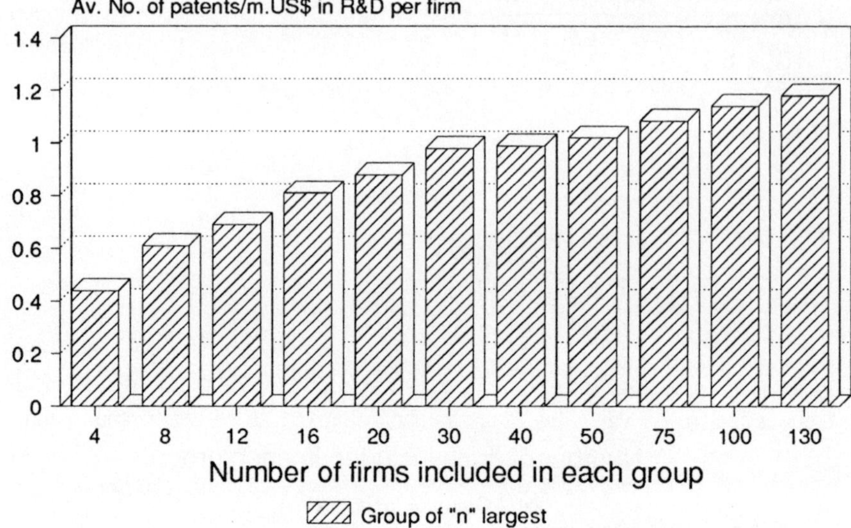

Figure 4.4   The inventive productivity of industrial R & D in the United States measured by patents (adapted from Soete, 1979).

4.4.2 Who innovates?

As demonstrated by Freeman (1982, Chapter 2), innovation has increasingly become an industrial and professional attribute. The growth of scientific content in its technology and the elevation of the R & D costs made innovation an industrial prerogative, snatching the role previously played by individual entrepreneurs.[16] Figure 4.3 (based upon data provided by Soete, 1979) exemplifies how concentrated is the research effort on the industry in general (the data is relative to the largest 130 American companies). As shown on the chart, the average investment in R & D in the four largest companies is just above six times the average among the whole group. Considering the capital intensiveness of the chemical industry, it can be expected to have an even greater level of concentration on R & D investment (besides invention productivity and, mainly, innovation productivity).

In terms of product innovation there is clear and unrefutable evidence of the virtual 'innovation oligopoly' composed of the 'heavy weights'. In analysing the innovation productivity in the pesticide segment, for example, Achilladelis *et al.* (1987) found that from 846 new products, introduced between 1930 and 1980, 381 came from only 10 companies. Even more important, they concluded that among the 85 major market successes 47 came from these 10 companies. Similar results can be found for synthetic materials. Freeman (1989) found that the five largest chemical companies were responsible for the introduction of 66 per cent of the major market successes in synthetic material from 1935 to 1950. Stobaugh found that all nine petrochemical commodities he researched were introduced to the market by large companies and attributed such a factor to their intrinsic capacity in terms of risk-pooling, better reputation, superior resources, wider scope and greater learning. Another piece of evidence comes from the SAPPHO project (see Rothwell *et al.*, 1974) which showed that research projects (in the chemical industry) with larger teams and investments have a larger percentage of success. In one of the most complete studies about innovation in the chemical industry, Achilladelis *et al.* (1990) found results which not only confirm SAPPHO's and Stobaugh's finds but go further, showing that radical innovations are usually more profitable than incremental ones. His results also show that the radical innovations used to become part of the 'corporate technological tradition'.[17] This 'tradition' acts as a 'force' to keep radical innovators in the innovated sector for long periods of time, subsequently introducing several minor innovations based upon the original ones.

In spite of all this evidence, however, there is considerable room for discussion about the role of smaller firms in terms of minor and process innovations. The work of Soete (1979) provides some support in favour of this discussion as shown in Figure 4.4. As can be seen from the chart, the average number of patents per millions of dollars expended in R & D is three times larger in the whole group of companies than it is among the four largest. Many reasons can be given for this higher productivity such as a stronger propensity for patenting (in this case the picture need not necessarily be interpreted as an indicator of invention productivity but, instead, of 'patent productivity'). It is very probable, however, that owing to their own limitations in terms of size, the smaller companies have a tendency to research for minor innovations (the ones that they can be more confident about accomplishing). One important contribution to this line of thought can be taken from Rocha (1984). This study shows an interesting comparison of technical performances between a large and very innovative company (ICI) and a small, mono-producer company (Polypropileno S/A) situated in a less-developed country (Brazil). Polypropileno built a plant for polypropylene production based on technology licensed by ICI. This plant was almost identical to the third ICI plant using that core technology. In spite of the largest (by far) technological capability of the licenser, the Brazilian company achieved, after a few years, a clearly better technical performance (measured in terms of several indices such as raw material and energy consumption per tonne of product). 'Very typically, the work involved the maximization of the equipment available through modifications on its operational conditions, recipes improvements and a number of minor technical changes' (p. 154).

Rocha highlighted that in spite of its proved capability in terms of small 'technical changes', Polypropileno didn't provide any major innovation to the process, while at the same time ICI was able to develop a radically new process.[18] It is important to observe that the genesis of radical innovations is not necessarily the beginning of a more profitable business. Hollander (1965) had showed that consecutive minor innovations can have an important role to play in a business's performances. In this classical work, he

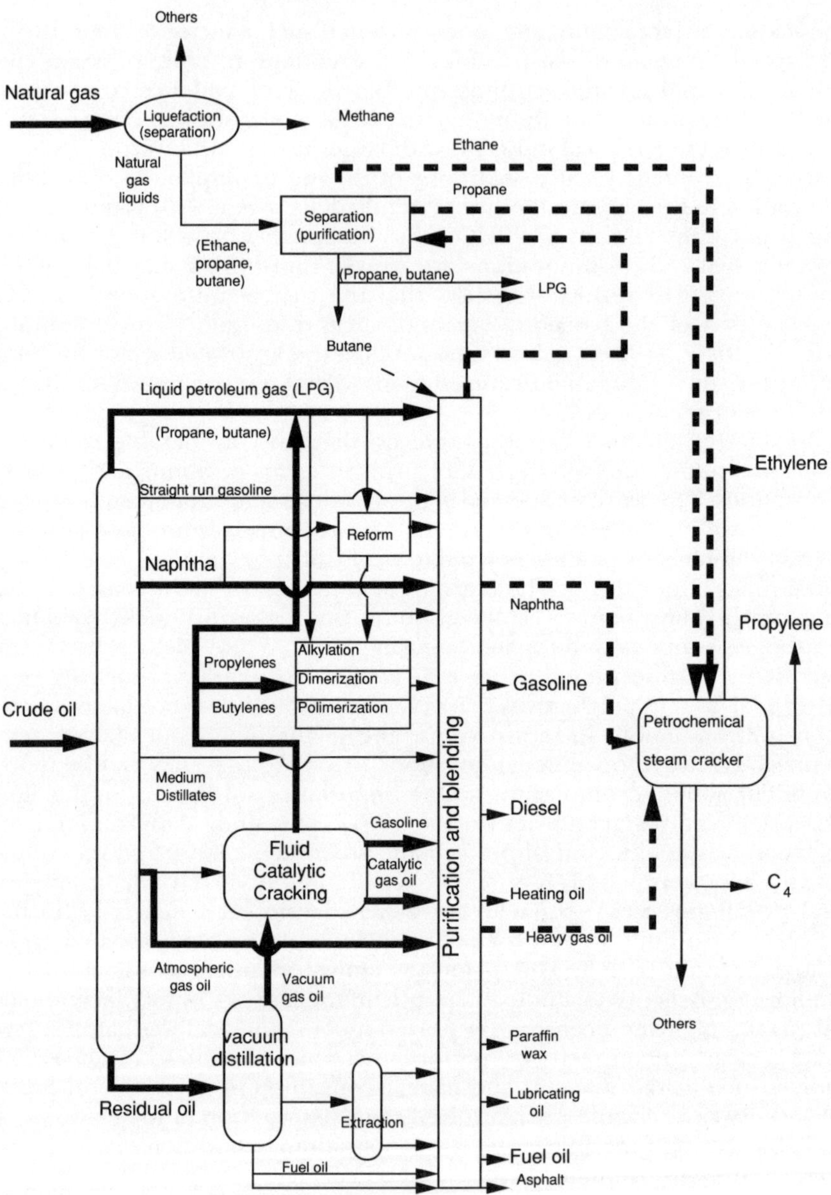

Figure 4.5 Simplified diagram of petrochemical feedstock production from crude oil and natural gas.

showed that the contribution of the several incremental innovations introduced by Du Pont in its rayon plants had a major importance on its productivity improvements, even more important then their radical innovations.

If there is, however, a part of the chemical industry which can be economically more sensitive to innovations, this part is the petrochemical feedstock. This subject is discussed below.

4.4.3 Petrochemical feedstock: a strategic point

As described above, olefins (particularly ethylene and propylene) constitute the main building blocks of the petrochemical industry. The two significant feedstocks for olefin production are naphtha and natural gas. The economics of olefin production and its connections with its feedstock is one of the most strategic aspects of this industry. Despite the huge complexity of the industry, this particular sector can be used to exemplify the type of problems, options and decision-making variables that it contains. It shows also the intrinsic linkage between this industry and the petroleum refining industry.

Figure 4.5 shows diagrammatically an oversimplified view of the main unitary processes involved in petrochemical feedstock production from crude oil and natural gas.

The oil refining consists of several unitary processes designed to break down the complex mixture of substances and chemical functions into its constituents.[19] Its main challenge is to match the market demands for its different products with the natural composition of the crude oil. In a simple refinery the quantity and variety of products is determined by the crude-oil composition. Few of these refineries are still in operation. Most refineries have unitary processes to crack the larger molecules into a mixture of smaller ones and also to combine different small molecules into larger ones. The main cracking process of interest for this section is 'Fluid Catalytic Cracking' (FCC). In terms of 'combining' processes, however, reforming, alkylation, dimerization and polymerization all have an important role to play.

The border (if there is one) between the petrochemical and oil industries is 'steam cracking'. This process is used to produce some of the main basic petrochemicals such as ethylene and propylene among others. As shown in Figure 4.5, the main feeds to this process are ethane, propane, naphtha and gas oils. The choice of one (or more) of these feedstocks is a major strategic decision. The seasonal characteristics of these products, however, add some more complexity to the equation as explained below.

Motor gasoline is the main oil-industry product for which demand usually exceeds the 'natural supply'; heavier products, mainly fuel oils, are present in a much higher proportion than their demand. In order to balance this equation, FCC is used (see Figure 4.5) to increase the refineries' yield of gasoline and producing by-products such as catalytic gas oil and liquefied petroleum gas (LPG). As the consumption of gasoline in the northern hemisphere is greater in summer than in winter, the FCC operation peaks in the early summer, fuelled by gasoline price rises. As a counterpart, gas oil and LPG prices fall as they are over-produced (see Longley, 1991). These variations in price and demand became more accentuated during the last few years following the phasing out of leaded gasoline which requires a larger amount of naphtha to be re-formed in order to produce the high-octane gasoline. In this situation, a petrochemical cracker wanting naphtha will need to bid above its premium value in gasoline production. On the other hand, when wanting LPG it will buy a likely surplus at low cost, (although the fractions butylenes, and eventually propylene, of this LPG prevenient from FCC will also be used through 'combination processes' to produce high-octane gasoline).

In the United States the volume of natural-gas production (and its composition) has since the 1960s provided enough ethane, and later propane, for the production of most of its ethylene. Also in the United States the high consumption of gasoline has prevented any excess of naphtha for cracking, making gas oil their most usual option for feedstock flexibility. In Europe owing to the smaller availabilities of domestic sources of natural gas and the low demand for gasoline (up to the 1970s), naphtha became the natural choice for cracking. Since the 1970s, however, European refineries have to build FCC units to keep pace with motor-gasoline demand, generating some surplus of LPG. European naphtha crackers have added LPG flexibility since the beginning of the 1980s. Also the discovery of the North Sea reserves has helped to change the situation. However, owing to the high costs involved in the transport and storage of natural gas,[20] it can be attractive only in Scotland, the north of England and Norway. In general, European crackers have a more limited flexibility than their American counterparts, generating wider seasonal price differentials. The condition of Japanese and other Asian producers in terms of petrochemical cracking is even more dependent upon naphtha than the Europeans and their flexibility smaller. As the Asian countries are experiencing a faster growth in petrochemicals, the demand for naphtha is expected to grow to a much tighter balance with production. This possibility helps to justify the industry's investments in research projects concerned with alternative feedstocks as described in Section 4.4. This situation is already stimulating fast changes in the Japanese industry, which intends to reduce its dependence upon naphtha from 95 per cent to 60 per cent by swapping this capacity with NGL and LPG (*Chemical Week*, 1992a).[21]

While all the discussion above is quite important to the present and near future of the chemical industry, it can be seen as one of its specific challenges. Although considering only its intrinsic problems is already a complex matter, the industry will inevitably also pass through some more generic challenges such as an eventual emergence of a new industrial 'techno-economic paradigm' as discussed below.

4.4.4 The emergence of a new 'techno-economic paradigm' and the chemical industry

The search for evidence of the emergence of a new techno-economic paradigm is among the most important studies in and out of the academy at the moment. This importance is proportional to the economic effects of

Table 4.4 Some of the main characteristics of the four 'Kondratiev waves' (adapted from Bessant, 1991).

Prosperity periods	Industrial revolution (1770-1820)	Victorian prosperity (1840-1870)	'Belle-epoque' (1890-1930)	Golden age (1950-1970)
Cluster-technologies	Early mechanisation	Steam power Railways	Electrical and heavy engineering	Mass production
'Carrier-industries'	Cotton Pig iron	Coal Transport	Steel	Energy (oil)
Embryonic industries (future carriers)	Steam-engines	Chemicals	Aluminium Synthetic materials	Computers automation
Limitations of growth	Limits of small scale manual processes and tools	Limits of steam power	Limits of iron usage	Limits of scale advantages
Managerial stereotypes	Individual entrepreneurs	Large companies	Giant multinational firms	Corporate divisionalizing

such a socio-technological phenomenon, which is closely connected with the economic waves of growth and decline first identified by Kondratiev. Some of the characteristics of the four long waves are depicted in Table 4.4 (see Bessant, 1991).

The use of the 'paradigm' concept received its first impulse with the work of Kuhn (1962), who described scientific progress as continuous incremental changes punctuated by periodic 'paradigm-changes', where major scientific discoveries changed the way through which the scientific community sees the world (for example, the advent of quantum mechanics). Dosi (1982) extended this concept to the technological paradigm which connects with changes in 'technological trajectories' (see Chapter 2). The broader discussion about 'techno-economic paradigms' was introduced by Perez (1983). In accordance with those ideas, a change in the techno-economic paradigm brings with it a whole range of new products and processes, involving major institutional changes in society since existing management systems at firm and governmental levels may be inappropriate for the promotion of the new technologies.

Bessant, who refers to this new paradigm as 'post-Fordist', noted that 'this paradigm is by no means clearly defined, but includes a number of features:

– greater emphasis on non-price factors;
– greater emphasis on flexibility in technology;
– greater emphasis on flexibility in organizational structure;
– changing relationships within and between organizations.'

In 1985, Perez had defined its main features, similarly, as:

– Information-intensity rather than energy-intensity in products and processes;
– Flexibility in process investment and in product mix rather than the dedicated capital equipment and standardized products range characteristic of the Fordist mass-production system;
– Systemation, not just automation, through the uses of information and communication technologies integrated to all major administrative functions within firms.

In describing the effects of the new paradigm upon the chemical industry Freeman (1989) wrote:

...chemical firms, like all others, are deeply affected by the capacity of computer technology to integrate various functions within the firm by the capacity of new communication technology to link together diverse production sites, R&D laboratories and other activities within the firm. Chemical engineering activities and specially the design function become more flexible and again permit more rapid changes in both products and processes... Thus the change in techno-economic paradigm does already deeply affect the chemical industry and will do so even more in the 1990's.

4.5 Conclusion

This short review of the chemical and petrochemical industries' evolution shows the crucial role played by innovation in their struggle for survival. In spite of their maturity, innovation will be kept as a central issue of any discussion about their future. No one other industry presents so complex a network of products, producers and processes. The fact that almost any chemical product can be made from several different raw materials each using several different processes will keep its dynamics and intrinsic science intensity as long as competition exists within these industries.

As shown throughout this chapter the pattern of innovation generation changed from the 1870s. Since that period it has become more and more dependent upon highly skilled professional research. During the last five decades this has become not only professional but also capital-intensive. Although, as shown above, it doesn't limit so strongly the capacity to produce minor innovations (mainly in terms of process), research is what often enables smaller companies to keep their competitiveness for a variable time span.

The present picture of the industry's main technological and strategic concerns can be conveniently broken into three parts: feedstocks, commodities and specialities.

– Petrochemical feedstocks are one of the areas of heaviest R & D investment at the present. Since the first oil shock the industry has turned its attention to the development of substitutes for naphtha. From a range originally consisting mainly of natural gas, coal, tar-sands, oil-shale and fermentation of molasses this search reduced to a single-component: natural gas. An analysis of the main names involved in heavy research into natural-gas processing shows a great majority of oil companies (see Section 4.4). This observation leaves no doubts that any major innovation will come from a large company or from an association of some of them.
– The field of chemical specialities shows clear signs of growth as a whole, pushed by digital techniques, flexible manufacturing systems and heavy investment in bio-science related areas.[22] All this development seems intimately related to the main features of the new paradigm as discussed above.
– In terms of petrochemical commodities, the oil companies have already a major share of the market, mainly in terms of bulk plastics. Among the largest five chemical companies (Du Pont, BASF, Bayer, Hoechst and ICI) there are clear moves towards specialities as a counterpart of their retrieval from the bulk chemicals. Among these five companies, Du Pont is the only one not investing heavily in pharmaceuticals.[23] Among the four European giants, however, the best results during 1990 and 1991 were achieved by Bayer, which has the fastest-growing pharmaceutical business in Europe. The large corporation's move from bulk chemicals towards specialities is a clear issue. This move is, however, a corporate strategy issue (see Chapter 1). What is not clear in terms of commodities is what type of strategic decisions and directions are made at the business level and what type of influence the eventual new paradigm can have upon the full range of companies and businesses operating in this field.

Notes

1. Up to 1860 (owing to the corrosive nature of soda ash) sodium carbonate was transported to the end-user where it was finally transformed into soda ash. With

the introduction of steel drums in 1860, this last step began to be done on-site.
2. The Solvay was the first continuous chemical process. It presented advantages in terms of raw materials and labour in relation to Leblanc's. The first plant using this process was started up in 1865.
3. In 1905, however, one of the main signs that British chemistry was still alive was through its pioneering of the man-made fibre sector. During that year Courtalds (then a silk weaving firm) innovated by introducing the industrial manufacture of rayon.
4. In terms of the range of boiling points, petroleum components can be divided (in a simplified view) respectively as: refinery gas (a mixture of methane, ethane, propane and butane), gasoline (a mixture of hydrocarbons containing from five to eight carbon atoms per molecule), kerosene (a mixture of hydrocarbons with 11–12 carbon atms per molecule), gas-oil (a mixture of hydrocarbons having from 13 to 25 carbon atoms). The residue is further 'broken' down into lubricating oil, paraffin waxes and asphalt.
5. The Dubbs process, whose pilot plant started in 1918, led to the formation of Universal Oil Products Company (UOP), an influential process design firm (see Achilladelis, 1975). The Cross process also led to the creation of a licensing company but, unlike the UOP, only lasted 20 years.
6. As result of the merging of British Dyestuffs, Brunner Mond, Nobel industries and United Alkali.
7. In order to keep its competitiveness, the original Houdry process was significantly altered to make it continuous, resulting in the Thermofor Catalytic Cracking (TCC), operated from 1944, and in the Houdryflow installed after 1950.
8. Up to that period the United States was the only country with a high consumption of gasoline, generating a huge amount of refining by-products at low cost.
9. The work of Dr Ziegler on catalytic polymerization, first deployed as a process development for polyethylene production.
10. Polymers such as polyethylene and polyacrylates at that time had almost exclusively military applications such as radar and airplane-cockpit coverture.
11. For a list of products composed of phenol, methanol, vinyl chloride, styrene, acrylonitrile, cyclohexane, isoprene, ortho and paraxylene, Stobaugh (1988, p. 127) observed a price decline of 68 per cent in real prices between 1961 and 1973.
12. The second major area of research is the so-called 'materials area' where those companies are investing mainly in high-value materials such as engineering plastics.
13. The paper cites other important innovations such as a new process for the production of acrylonitrile from propane (extensively discussed in Chapter 5) and a direct oxidation process for propylene oxide manufacturing.
14. Natural gases, depending on their location, can have from 69–96 per cent of methane and variable amounts of ethane, propane, butane and others. In the United States it has been used as petrochemical feedstock for decades (but only the fractions of ethane, propane and butane, collectively known as natural-gas liquids – NGL). The present research effort is concerned with the use of methane once it is the main constituent.
15. If we break down the classification of commodities into true and pseudo-commodities a similar pattern can be seen, again, with pseudo-commodities more disposed to product innovation. On the other hand, a comparison between fine-chemicals and specialities cannot be expected to show great differences.

16. In spite of such changes in the technology nature, there are several nuances to be considered in a broader analysis. Katz (1972), for example, showed that in Argentina the technical progress of a comprehensive group of industries was strongly associated with 'adaptation or copy' of technology and that this practice was more associated with other 'technical departments' than with R & D departments.

17. Achilladelis makes a parallel between this concept of 'corporate technological tradition' with that developed by Pavitt (1984) of 'knowledge accumulation'.

18. Rocha speculates that ICI could have moved its best people to the development of the new process, not paying the same attention to the old process as did Polipropileno. It is important, however, to highlight that it was an operational unit and certainly was not left 'adrift' in a highly competitive market like the one for polypropylene.

19. It is important to emphasize that one of the characteristics which differenciates this industry from the petrochemical one is the fact that most of its products and streams are still a mixture of several different substances (although much more homogeneous than crude oil), while in the latter most of its products are almost pure products (typically above 95 per cent, usually above 98 per cent or 99 per cent).

20. While crude oil and naphtha are easily transportable (with costs usually under US$10 per tonne), natural gas requires liquefaction, terminals, shipping and receiving conceived as a whole in order to be explored. Such an enterprise can only be backed by long-term contracts and at a cost usually over US$100 per tonne (Longley, 1991).

21. It is important to observe that Japan is already an important importer of NGL (which has been used as a 'clean' raw material in the production of their gasoline as a consequence of strict environmental regulations). This combination of demands certainly contributes to the essential economy of scale of the whole system.

22. In spite of such an impetus, several individual enterprises are resulting in major losses: for example the ICI advanced plastics failure (see *Financial Times*, 1991b).

23. Different from the other 'big-five', Du Pont is one of the most back-integrated companies, with considerable interests in oil (through Conoco's oil business acquired in 1981).

5 Case-study A (acrylonitrile producer)

5.1 Introduction

Case A is a strategic business unit (SBU) within the Chemical Division of a very large oil company, which will be referred to as 'Oilcomp' (The real names of companies, divisions and businesses will not be used in this book following agreements on confidentiality). This SBU has its headquarters in the United States and is responsible for the production, sales, research and development of three main products and a handful of their by-products and derivatives. Among these three main products is an intermediate petrochemical called acrylonitrile.

Case A produces approximately 15 per cent of the world's total acrylonitrile output which means about 500,000 tonnes a year. The most important data, however, are related to its technological leadership, as more than 95 per cent of world production is made using their technology. Such technological leadership is very unusual for an intermediate petrochemical product. Perhaps still more interesting is the difference between the two percentages. While 95 per cent of world production depends on their technology (propylene ammoxidation), only 15 per cent of this is made at their own plants.

Before discussing the reasons for such a discrepancy, it is important to discuss the take-over of 'SO-Corporate' by Oilcomp. SO-Corporate was the innovator company which in 1959–1960 commissioned the first plant using this revolutionary technology (propylene ammoxidation). This event is illustrated in Figure 5.1 where the main steps of the take-over are shown.

The explanation for such a big difference between the percentage of acrylonitrile produced through their technology and that produced through their own plants goes back to the 1960s when the company had just exchanged 25 per cent of its equity for a 50 per cent participation by Oilcomp in the biggest oil field in the United States (Prudhoe Bay in Alaska). In order to explore such an asset it was necessary to construct an

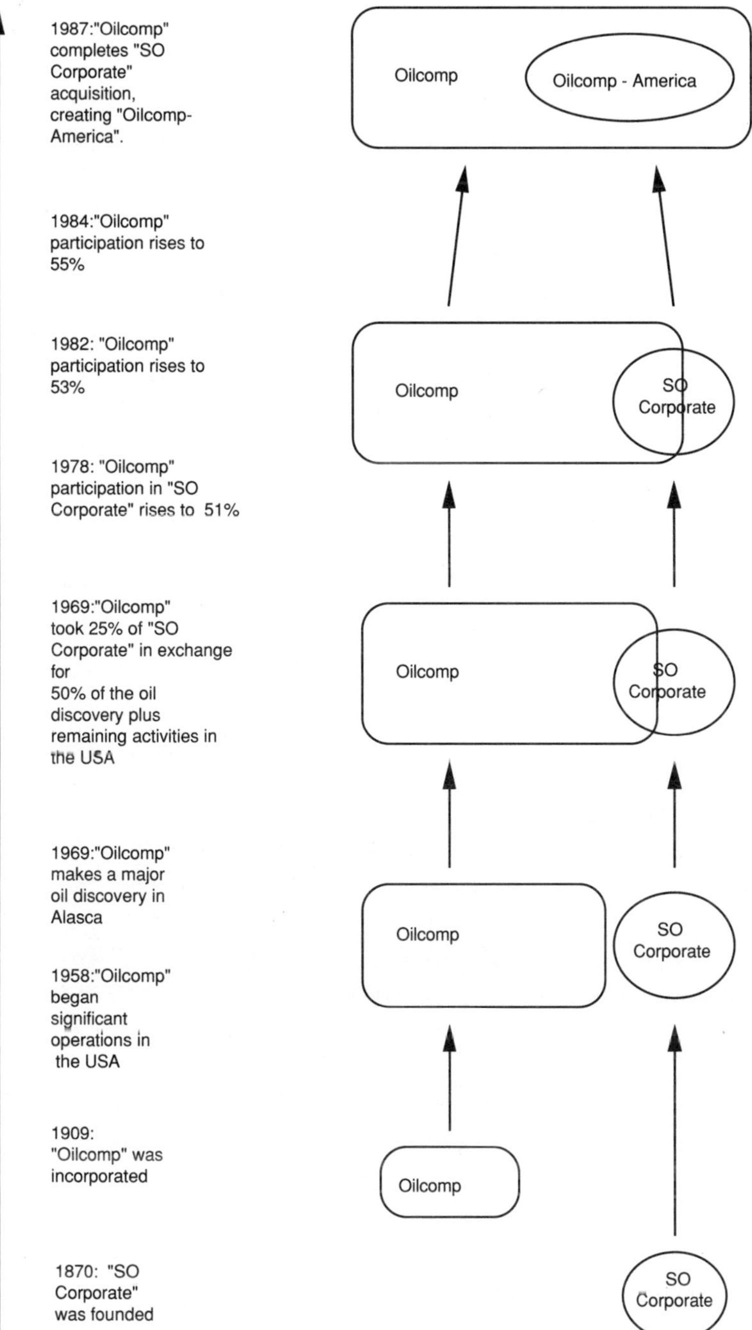

Figure 5.1   Main steps in SO-Corporate's take-over by Oilcomp.

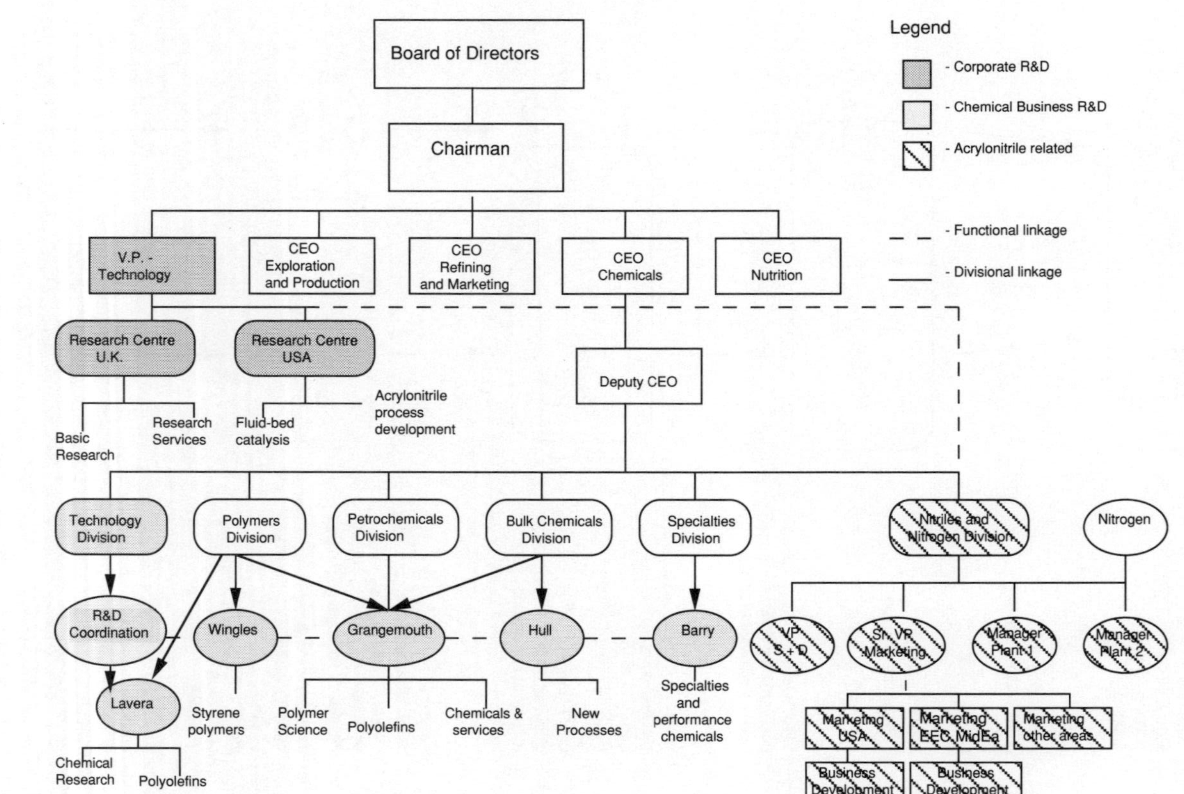

Figure 5.2 Outline of acrylonitrile and R & D related units in Oilcomp's structure.

96

800-mile Trans-Alaska Pipeline (a huge effort even for oil companies). The need for capital for the pipeline led SO-Corporate to license its newly developed technology for acrylonitrile production throughout the world. This technology has been licensed by more than 30 companies in 20 different countries. Stobaugh (1988) estimated that the company has earned more than US\$800 million from the commercialization of the product as well as the licensing of the technology. This strategy explains the chronic surplus of product, discussed in the market review section.

While the money was helpful for the construction of the pipeline it was not sufficient to avoid the take-over from Oilcomp in 1987. Presently, Case A has two fully owned plants in operation (both in the United States) and a third one, where Case A has 50 per cent of the shares, operates in Germany. During the execution of this research they shut down a fourth unit in Scotland.

As is the case with most of petrochemical businesses they have a global strategy, with departments throughout the world. All regional activities are, however, linked through a business structure, representing a matrix-type organization. Part of this structure (that of most direct interest to the objectives of this book) is shown in Figure 5.2.

5.2 'Oilcomp's' strategy analysis

From all the information collected during the interviews in Case A itself, as well as the information generated from interviews with other companies, consultancy firms and a literature review, it is possible to understand and describe its generic business strategy as well as its technology strategy.

In order to conduct this assessment, it is necessary first to obtain a generic understanding of the corporation's overall strategies. Oilcomp has defined its general strategy during the second half of the last decade as a continuous involvement with business in which they can keep 'selective excellence': that is, being among the very best; and 'critical mass', which means being of sufficient size to compete strongly in the market' (*Annual Report* 1987, p. 3).[1] In practical terms, it means that they intend to concentrate their efforts in traditional areas of business, where they already have a strong merchant position as well as a world-class technology. The implementation of this strategy can be seen through examination of the main business sales and purchases made within the cited time span. During this period Oilcomp divested some huge businesses such as Oilcomp Minerals (7 per cent of their assets), 1989; large part of Oilcomp Coal, 1989. At the same time they also made huge investments towards their strategy such as: the take-over of SO-Corporate, 1987; purchase of 10.7 per cent of their own shares from a major shareholder, 1987; purchase of Oil B, 1988.

This strategy reflected down-stream on the different businesses, as stated by the Chief Executive Officer from their Chemical division (*Annual Report* 1990, p. 21): 'We are continuing to focus our investments on those business sectors where we enjoy a combination of strengths – a leading technology, access to cost-competitive feedstock and an established market position.'

Table 5.1 Main acrylonitrile producers and their capacities (thousands of tonnes per year).

Case A	740[a]
Sterling Chemicals	275[b]
Russian producers	260
Asahi	250
BASF	230
Monsanto	220
DSM	180
Enichem	180
Chinese producers	175
Pemex	174
CPDC	170
Du Pont	170
American Cyanamid	160
Repsol	110
Hoechst	105
Romanian producer	100
Mitsubishi	100
Nitto	100
Acrinor	78
Petkim	70

a – Including its participation in other companies' production and shares
b – Disregarding other companies' shares in its production

Such a strategy can be clearly observed also in several smaller divestments : PVC (polyvinylchloride), 1982; chlorine, 1983; inorganic in general, 1984; 'Bakte' (United Kingdom), phenol and cumene, 1985; 'Bakte' (Asia and Singapore), others, 1986; hydroxyethyl cellulose, 1988.

In their strategy towards concentration, the Chemical division also made some investments such as: 'Chemicomp' participation in the market of LDPE (low-density polyethylene), 1987; 'Aclon' (detergents) and 'Tolco' (composites), 1988; construction of two new plants for LDPE production, 1989. Following this 'strategic down-stream wave', Case A also made strategic movements such as: shut down its small, obsolescent plant for acrylonitrile production in Grangemouth (Scotland), 1990; expanded its acrylonitrile plant at Greenlake (USA), 1989; commissioned a new acetic acid/anhydride, 1989.

A clear result of the strategy is that now the corporation is concentrated in only four divisions: 'Upstream oil production', 'Downstream oil production', 'Nutrition'[2] and 'Chemicals'. In this last division they are concentrated in three strong areas: 'Petrochemicals and Polymers', 'Acetyls and Nitriles' and 'Specialty Products'. Within Acetyls and Nitriles is the Nitrogen and Nitriles division. Sections 5.4 to 5.8 will focus on this SBU.

Table 5.2   Major acrylonitrile consumers (per product) and their consumption
capacities including subsidiaries (in thousands of tonnes per year).

Acrylic-fibre producers (.95 Kg of AN per Kg of fibre)

Montefibre	288
Bayer	160
Russian producers	160
Chinese producers	160
Case B	156
AKSA	130
Hanil	115
Monsanto	110

Adiponitrile producers (1.15 Kg of AN per Kg of adiponitrile)

Du Pont	540
Monsanto	240
BASF	170

ABS/SAN producers (0.27 Kg of AN per Kg of ABS/SAN)

GE Plastics	160
Dow	155
Chi Mei	150

NB Polymers producers (0.3 Kg of AN per Kg of polymer)

Bayer	26
Zeon	24
Case D	1.5

Acrylamide Producers (0.76 Kg of AN per Kg of acrylamide)

Dow	40
Allied Colloids	34

5.3 'Oilcomp's' R & D structure and strategy

Oilcomp global R & D expenditure (including both corporate and business) in the last five years has been around 1 per cent of its turnover,
almost £330 million in 1990. Between researchers and support staff, it
employs around 4,000 people.

Most of this R & D effort is conducted through its strategic units in their
own R & D laboratories (see Figure 5.2), employing around 800 people. A
second and much smaller part of this business unit's research is conducted
at some of its production sites, where almost all effort can be classified as
applied process research but where sometimes product and basic research
also happen.

As shown in Figure 5.2 there are five 'business-orientated laboratories'
situated in Lavera and Wingles in France; Grangemouth, Hull and Barry in
the United Kingdom. The first employs approximately 260 staff whose
main activities involve the development of products and processes for the
diversification of their range of ethylene oxide derivatives and also
polyolefins catalysis and products. At the Wingles' laboratory there are

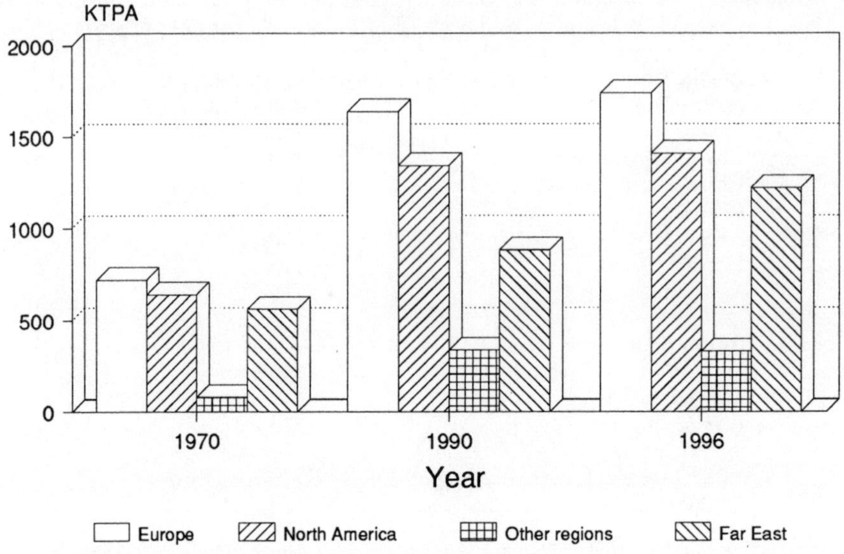

Figure 5.3 Acrylonitrile capacity per region.

approximately 50 researchers working solely on styrene projects (within the polymers division).

Grangemouth laboratory's 185 staff are divided between polymer science research (mainly polyolefins) and chemicals, (where R & D for a number of Chemical divisions chemical and petrochemical processes are carried out). At Hull 200 researchers are involved in new process development within the 'Bulk Chemicals Division'. Finally, at Barry laboratory 120 staff are directly involved in research related to specialities and performance products.

At corporate level there is long-term research, mainly not directly related towards any of the core businesses (although there are exceptions as will be described below). Interestingly, all the analytical development[3] is also at corporate level. The corporate R & D division operates two laboratories, the larger (in the United Kingdom), employing almost 2,000 people; the second one in the United States employs approximately 1,000 people. In spite of their corporate control, much of their activity is related to business R & D programmes. The Chemical Division's programmes alone (which include Case A), for example, employ around 600 staff from that corporate laboratory.

Conversely, the engineering support is split between corporate and business levels. Until two years ago the engineering department was almost completely at corporate level and was composed of three parts: unitary operations/equipments, processes and project management. Presently, all process personnel is spread throughout their SBUs, while most unitary operations/equipment and project management people remain at the corporate level.

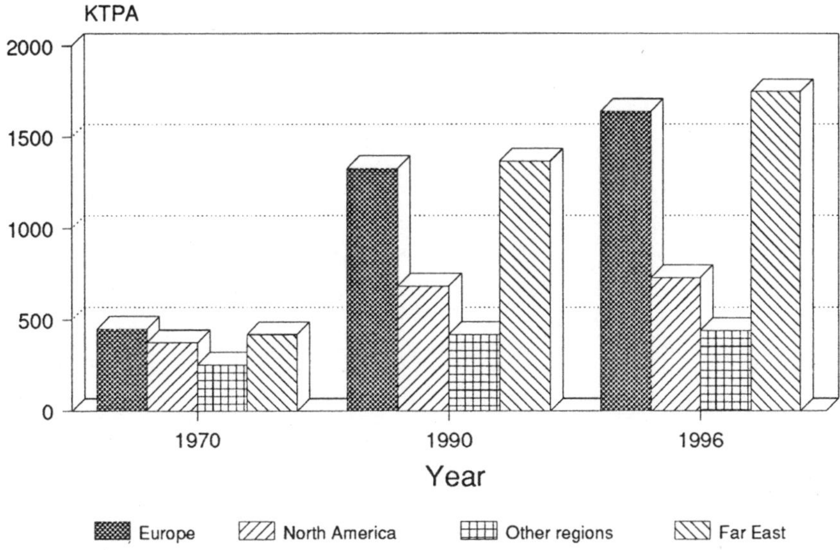

Figure 5.4   Acrylonitrile consumption per region.

5.4 Acrylonitrile market review

Several important aspects of acrylonitrile's market (such as the share of its consumption by each of the main derivatives) have already been discussed in Chapter 3. Table 5.1 shows the production capacity of the major acrylonitrile producers, while Table 5.2 indicates the capacities of the main consumers.

Owing to over-expansion from 1970 to the early 1980s, the market for AN has been chronically oversupplied. From 1970 to 1990 world consumption grew from 1,500 KTPA (Kilotonnes per year) to 3,800 KTPA, while production capacity increased from 2,000 to 4,200 KTPA owing to some new plants and 'de-bottleneckings' in all plants where this was possible. During this same period the distribution of world capacity and consumption had evolved in accordance with the charts shown in Figures 5.3 and 5.4.

Figure 5.3 shows AN installed capacity by region in 1970, 1990 and the accepted forecasting for 1996.[4] Figure 5.4 was developed identically to Figure 5.3 but considers the regional consumption of AN. Figure 5.5 is derived from the two previous charts and shows the surplus per region. Examination of these three charts makes it clear that:

– New capacity will be required in the first half of this decade;
– American participation in world consumption is dropping steadily;
– Far East consumption is growing sharply;
– Far East debit is growing sharply and steadily.

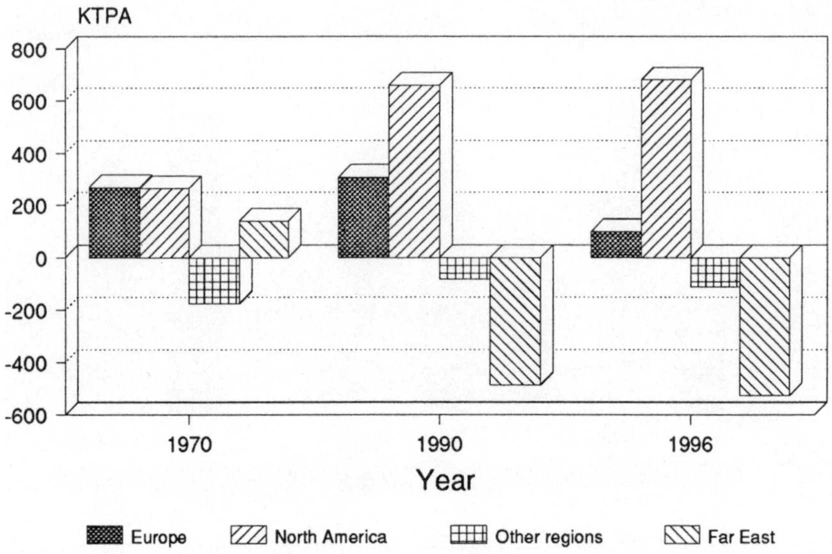

Figure 5.5 Acrylonitrile surplus per region.

5.5 Acrylonitrile technology review

Since the beginning of its industrial production, the technology for acrylonitrile manufacturing passed through different phases, each of them with one dominant process as discussed in more detail in Section 5.7.1. The last two phases were dominated respectively by the acetylene process and presently by the propylene ammoxidation technology. All world production is, currently, based on the propylene process which is described below.

The process for the manufacture of acrylonitrile is a typical petrochemical one. It consists of three main steps: reaction, recovering and purification, besides secondary units involving waste-treating and elimination.

The reaction takes place in a fluid-bed catalyst reactor where propylene and ammonia are admitted and reacted with oxygen from the atmosphere at a temperature of approximately 440° centigrade and 10 pounds per square inch (PSIG). The heat generated by the reaction is removed from the catalyst bed by serpentines and used in other sections of the plant. Reaction products (which involve a mixture including acrylonitrile, water, hydrogen cyanide, acetonitrile and others) are then cooled down in a heat exchanger and passed through a quench column where unreacted ammonia is neutralized by the addition of sulphuric acid to produce ammonium sulphate (one of the main effluents). The neutralized mixture (still in gasphase) is further cooled down and the non-condensed gases absorbed. The two currents are then rejoined and sent to the recovery section as shown in Figure 5.6.

In the recovery section the first step is an extractive distillation which

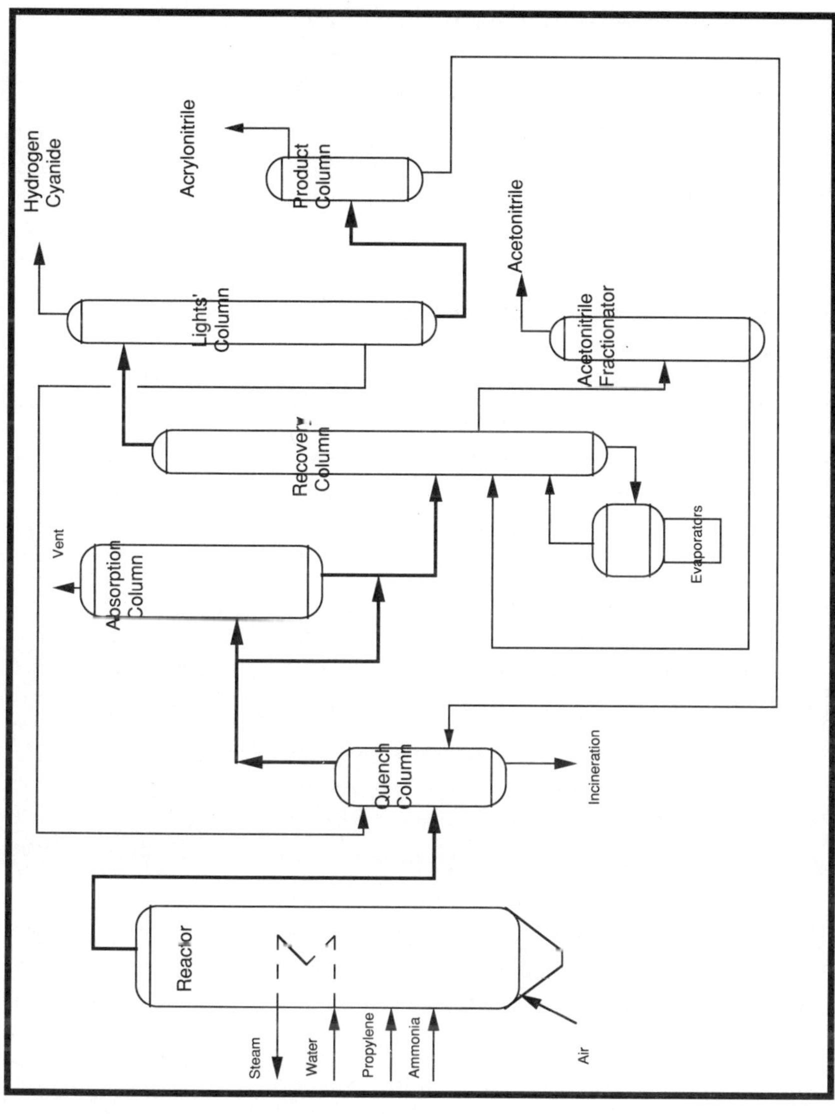

Figure 5.6 Case A's acrylonitrile process.

results in one stream of acetonitrile in water (contaminated with hydrogen cyanide) and the main stream containing acrylonitrile and hydrogen cyanide. The current of acetonitrile is distilled producing a solution of acetonitrile (which can be further purified or incinerated) and recycling the other stream back to the extractive distillation. The main stream coming out of the recovery column is then conducted to the purification section.

The purification is conducted first through the separation of the hydrogen cyanide (distillation) from the main stream (now containing nearly only acrylonitrile and water). The hydrogen cyanide (HCN) current is condensed and stored (usually sold). The main stream goes through a further distillation resulting in commercial acrylonitrile and a stream of water and HCN, which is recycled to the quench column.

5.6 Case A's business strategy analysis

As discussed previously, this business-strategy assessment will be based upon Porter's (1980, 1985) taxonomy of generic strategies, analysed throughout Chapters 1, 2 and 3.

Case A has most of its present production based in the United States; nevertheless, they act through a world-wide geographical network. Their marketing department is composed of four areas: North America, Europe and the Middle East, South and Central America, the Far East and Pacific. Each of these areas has its own personnel in marketing, business development and management services. Such a broad scope eliminates from consideration Porter's classifications based upon the generic label 'focus'.

The remaining possibilities (cost leadership and product differentiation) are quite different and easily identifiable. Considering Case A's technological advantage and the size of its plants (see Table 5.1), it could be possible that they operate through a cost-leadership strategy. In considering the type of product, a petrochemical commodity, such a conclusion would seem even more probable. Accordingly to interviews conducted, however, their generic strategy is 'differentiation'. In spite of their technological and, in fact, cost-leadership, their prices are comparable to but slightly higher than those of their competitors.[5]

Case A strives to differentiate itself in three ways: (a) technical assistance, (b) higher quality and (c) reliability and responsiveness.

(a) Technical assistance: the first factor is not a major difference, since the type of support Case A provides to its customers are of two main types – 'handling and safety' and engineering. Most producers can provide the first type of support, and in all likelihood, the other big producers at least can provide the same type of engineering support. For these reasons their advantage, although present, does not seem of significant importance.
(b) Higher quality: in considering the aspect of quality, it is necessary first to consider the type of product under analysis. As described before,

Table 5.3 Typical specifications of acrylonitrile.

Acrylonitrile	99% Min.
Acetone	300 ppm*/wt Max.
Acetonitrile	500 ppm/wt Max.
Acrolein	10 ppm/wt Max.
Acidity (as acetic acid)	35 ppm/wt Max.
Aldehydes (as acetaldehyde)	50 ppm/wt Max.
Copper	0.2 ppm/wt Max.
Hydrocyanic acid	10 ppm/wt Max.
Iron	0.2 ppm/wt Max.
Water	0.5% (wt) Max.
Peroxides	0.2 ppm/wt Max.
Appearance	Clear and free of suspended matter
Colour (APHA)	10 Max.
Distillation range	74.5–79°C.
Hydroquinone monomethyl ether	35–45 ppm/wt.
Stability (oxygen)	4 hs Min.
Ph	6.0–9.0.
Refractive Index	1.3882–1.292.
Titration value	2 Max.
Specific gravity (25/15.5)	0.7990–0.8020.

* parts per million

acrylonitrile is a petrochemical commodity. Basically this type of product can be bought from any producer throughout the world with the same composition and, in principle, the same quality. In the case of acrylonitrile, the specification accepted world-wide is like the one shown in Table 5.3.

Case A's product is in accordance with the same specifications used by almost all other acrylonitrile producers. The main difference is related to a treatment in which the product is passed through some ion-exchange resins. Case A claims to 'extra-purify' its product through this system. As a consequence, their product could have some narrower specifications, mainly relating to 'appearance' and oxazol content. These characteristics make it more appropriate for certain types of applications (certain ABS grades and one or two acrylic-fibre processes for example). However, such dependence upon an 'extra-pure' product is much more the exception than rule. Among the main consumers who rely upon such characteristics is 'Fibrecomp', who also constitute one of the case-studies in this book (See Case B, Chapter 6). Fibrecomp, owing to its relatively high consumption needs and on account of other reasons described in the next chapter, has its own ion-exchange system, which allows it to buy, in principle, from any producer in the world independent of its quality. Another factor relating to Case A quality is the adoption, three years ago, of a 'total quality' manufacturing approach.

In spite of all these factors, the combination of the technical capability of consumers (like 'Fibrecomp') with the existence of other 'high-

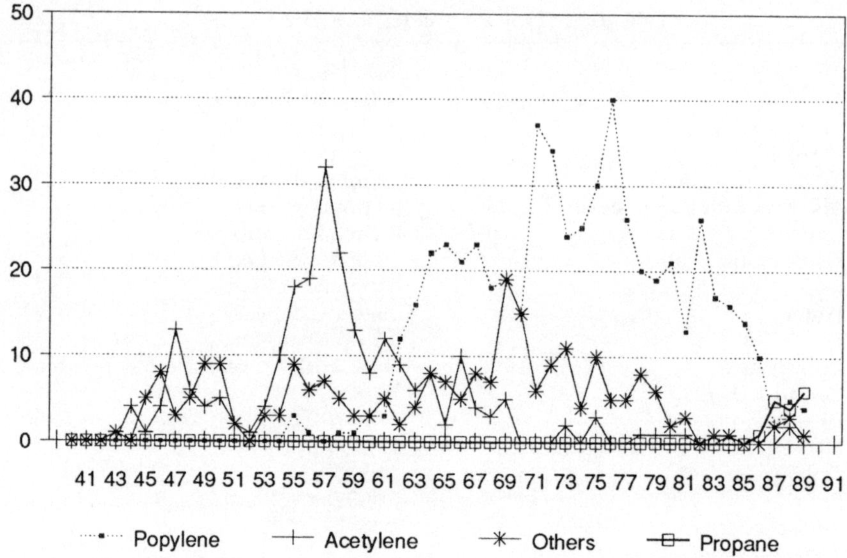

Figure 5.7 Acrylonitrile production-process patents.

quality' producers does not confirm this 'Case A's advantage' as their most important.

(c) Reliability and responsiveness: the main characteristic that determines Case A's product differentiation seems to be linked to its position as the largest producer and, most importantly, largest merchant supplier of acrylonitrile (they have only a small captive consumption, a different position from all other big producers). These data are shown in Tables 5.1 and 5.2, where most of the world's production and consumption is listed by company.

Case A produces acrylonitrile from six different reactors and two continents (including both their own reactors and others where they have capacity rights). In having the production based in so many different reactors, it is much less probable that some major breakdown could interfere with their production. This condition is reinforced by their supplying contracts with other producers which allow Case A to get large amounts of acrylonitrile from other producers in case of a discontinuity in their production. Also, their size and global presence makes it possible to respond to customers' requirements faster (in a global base), creating a major differentiation.

All these factors together make it possible to give their consumers a more effective guarantee of constant supply. This guarantee is not so possible for the other producers and is a considerable advantage considering the capital intensity of the companies down-stream from acrylonitrile. The existence of this advantage, together with the minor ones described above, explain and confirm their differentiation strategy.

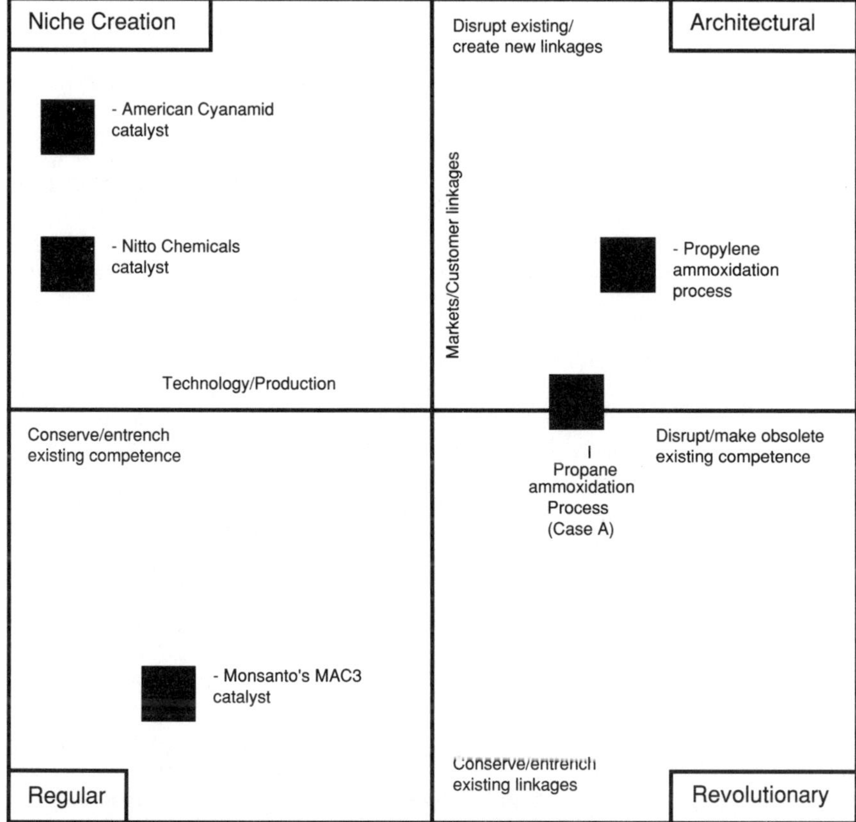

Figure 5.8   Positioning acrylonitrile process innovations in the Abernathy and Clark chart.[8]

5.7 Case A's technology strategy analysis

Using the data from Section 5.5, referring to technology review and Section 5.4, referring to marketing review, this section is intended as a discussion of Case A's technology strategy through the three-level strategy analysis proposed by Mulder and Vergragt.

5.7.1 Macro-level

Mulder and Vergragt identified the 'macro-level strategy' in their survey of companies accounting for a common part of their technology strategy. Theoretical support comes from economic theory, mainly from the contributions of Nelson and Winter (1977), Abernathy and Clark (1985) and Dosi (1982).

Section 5.5 discusses the evolution of the process technology for acrylonitrile production. This evolution can be divided into three main steps:

– Acetylene process development,
– Propylene ammoxidation process development; and more recently,
– Propane ammoxidation development.

Figure 5.7 shows the number of patents issued by year for each of these processes.[6] These data were drawn from the *Chemical Abstracts* (the main abstracting work for the chemical literature).

As shown by Achilladelis (1973), acrylonitrile's industrial production began (1938) with a process using ethylene oxide and hydrocyanic acid. This process became obsolete in 1945 following the introduction of the acetylene process. The acetylene route, in spite of its intrinsic economic inefficiency, had created a new 'technology/production competence' and at the same time a new 'market/customer linkage', characterizing an 'architectural innovation' in accordance with Abernathy and Clark's (1985) taxonomy of innovations (see Figure 5.8). With this new process the price of acrylonitrile dropped from 1.8 to 1.0 RM/KG[7] (Achilladelis, p. 87).

The introduction by Case A of a propylene ammoxidation route in 1960 made acrylonitrile prices drop again from US$0.23/Kg to US$0.15/Kg. Such a difference led ICI, for example, immediately to shut down its own plant and import acrylonitrile from the United States. This innovation evidently disrupted the previous 'technology/production competence' and made the derivatives (mainly acrylic fibres and ABS/SAN resins) much more competitive, enlarging the market and also disrupting 'market/customer linkages' (see Figure 5.8).

The fact that more than 95 per cent of the world production of acrylonitrile is based in Case A and its licensees' production is a clear demonstration of this innovation success. It is also clearly a 'natural technological trajectory' towards processes using cheaper raw materials and more sophisticated catalysts.

The patents survey (condensed in Figure 5.7) shows that almost all other patents issued for the acrylonitrile (from 1960 until 1985) production process were incremental innovations within the same technological trajectory. This technological trajectory is confirmed by Case A's main technological competitors such as Nitto Chemicals in Japan. This company developed a successful catalyst which allows higher yields from the main raw material (propylene) to acrylonitrile than the Case A original catalyst. It reduces the production of by-products (hydrogen cyanide and acetonitrile). In spite of the success of this catalyst, it was developed to be used for the same process and equipment developed by Case A. This type of innovation is within Abernathy and Clark's classification of 'regular' innovations. Another incremental innovation had been made by American Cyanamid which developed proprietary modifications on 'Case A catalyst 41' in order to improve the yield of hydrogen cyanide, which is used captively as a raw material in other processes. This second catalyst can be classified as 'niche-creation' innovation as it creates a new opportunity but in very specific cases (where it is beneficial to increase yields to by-products at the expense of AN output).

5.7.2 Meso-level

Before beginning the strategy analysis, it is interesting to mention that almost all of Case A's R & D is conducted in the American corporate laboratory. The reasons for such an unusual course seem to be mainly historical, owing to the relatively short time since the SO-Corporate acquisition, and can be associated with Achilladelis's concept of 'Corporate Technological Traditions' described in Chapter 5. On the other hand, a great part of their effort was directed towards the development of a catalyst that promotes the production of acrylonitrile by a one-step ammoxidation of propane. Such a breakthrough is absolutely impossible without heavy long-term basic research in catalysis.

The meso-level is at the level of interacting company strategies, accounting for opportunities which are circumstantial 'to the organization, markets and socio-political environment of companies'. As described in the sections referring to methodology, most of the data used here for meso-level strategy analysis are extracted from producing and consultancy company information, complemented by interviews with Case A.

The Case A process for acrylonitrile (or AN) production was so successful that ten years after (1970) its first plant was started, 75 per cent of world output was already based upon their technology. As described previously, Case A's share of the world acrylonitrile production is not as large as its share in terms of production technology. However, Case A's business strategy is heavily based upon their logistic. As described in Section 5.6 their differentiation strategy depends not just on the guarantee of constant supply but also on the speed of response. The production and consumption data presented in the market review above show a dramatic increase in the importance of the Asiatic market. These data, together with the virtual impossibility of new significant 'de-bottleneckings', leads to the conclusion that the new capacity could come from new plants in the Far East and, consequently, that this region should be the logical location for new investment. In order to keep its leadership in terms of supply (as required by its strategy), Case A should make such investment. The main problem is the absence of propylene capacity in the region. Presently propylene costs up to 25 per cent more in Korea, for example, than in Europe. This barrier provides the opportunity to enhance Case A's differentiation and (as shown below) to test some of the models discussed in Chapter 3.

In Chapter 3 a model was derived from Wilkinson's (1987) work, and the 'modification of process to utilize alternative raw-materials' was suggested as an R & D counterpart to suppliers' strength. Such a counterpart is exactly the action taken by Case A. In accord with Dirks (1991) and Oilcomp *Annual Reports* (1988, 1989, 1990), Case A is presently running a pilot plant for AN production from catalytic ammoxidation of propane. It is important to point out that propane is an abundant and cheap raw material in the Far East.

5.7.3 Micro-level

The micro-level, defined but not analysed by Mulder and Vergragt, will be analysed from interviews conducted in Case A, as well as from dozens of internal publications such as annual reports and reviews obtained directly from Oilcomp and Case A.

Case A's approach to the linkage between technology and business strategy is conducted from two sides: the R & D and the marketing departments. As shown in Figure 5.2, inside each of the regional marketing departments is a unit called 'business development'. The business development unit is formally in charge of the linkage between marketing people and the R & D division.

People in the business development units have a technical background and are kept well updated on the R & D progress in possible areas of interest. This updating is made via monthly R & D reports, as well as frequent visits and seminars. As an example of the type of participation from business development people in R & D activity (according to interviews conducted inside this unit), it is a usual activity, for example, to ask for higher or lower priorities for different projects in accordance with its business perspectives. This type of 'interference' is clearly within the R & D strategic space as defined by Coombs *et al.* (1987, Chapter 3). The existence of the business development unit and its responsibility for defining priorities in R & D projects characterizes a formal approach to the linkage.

On the other hand, these business development units do not use any specific tool or methodology for priority setting of the R & D projects other than classical business strategic aids (such as BCG matrices) and economic evaluations (such as present net value). According to the interviews conducted, for example, a typical type of request from the business development unit to the research staff is: 'We need 1 per cent higher yield per loop of reaction in order to have a competitive advantage in relation to the competitors' technology.'

5.8 Conclusion

The conclusions to be described here will be only those that relate to this specific case-study. The overall conclusions obtained from the examination of all the case-studies will be presented in Chapter 10.

The corporate strategy of 'Selective Excellence' stated in the Oilcomp annual report of 1987 (and apparently 'deliberated' since 1985) seems to have been effective.[9] Nevertheless, it is not possible (or even necessary) to assess how much of it was 'deliberate' and how much 'emerged'. Certainly it was partially imposed by the world's economic situation after the second oil shock. Contributing to this conclusion about its effectiveness is the observed Chemical Division alignment with the corporate strategy. Even Case A's technological strategy towards differentiation seems aligned to the defined corporate strategy. Such conclusions indicate that it is both interesting and important to assess the corporate strategies before the analysis of technology strategies (in this case-study they are clearly influencing each other).

Case A's business strategy can be easily identified with Porter's generic strategies' taxonomy. Its classification as 'differentiation' is a surprise (for reasons discussed above in Section 5.6) but is consistent with the corporate strategy as well as with the technology strategy as it is directed towards its very essence as the technological and merchandise leader. Its implementation reinforces the requisites of 'critical mass' and 'technological leadership' as stated four years ago by the chairman.

The organizational response to the strategies through the direct subordination of most of the R & D effort (and laboratories) directly to each business could again be an indication of alignment with the corporate and business strategies.

The use of Mulder and Vergragt's (1990) framework for technology-strategy assessment seems useful. At the macro-level it was possible to identify the 'natural trajectory' of acrylonitrile production towards the development of processes using cheaper raw materials. The position of some of the innovations (within that trajectory) in the Abernathy and Clark (1985) model reinforces the general vision of the innovations and leads to the discussion of the role of the propane ammoxidation route as a potential 'architectural' or 'revolutionary' innovation. Such expectations can be explained by comparing the effects of the cost reduction which occurred when the propylene process was introduced with what might happen if the propane process were to bring about a similar reduction in price.

Although it can not be affirmed that the development of propane ammoxidation is a consequence of their 'differentiation generic strategy', it is clear that it will reinforce such a strategy. The technology for propane ammoxidation will make it possible for Case A to 'choose the raw material' to be used in its new investments in acrylonitrile (if not during this decade almost certainly for the first plant to be built in the next decade). It seems a very important advantage, particularly if it is going to be built in the Far East (as suggested by the market review presented above). Such an investment will spread Case A's production even more, helping not only to strengthen its advantages in terms of guarantee of supply and responsiveness but also lowering its costs even more. It is also important to emphasize that such strategy may be seen as a consequence of their corporate technological traditions acting upon a mature product (and consequently imposing some 'emergent character' to their strategies). It should be noted, as well, that propane is easily available for an oil company such as Oilcomp, while propylene production usually requires a petrochemical cracker (see Figure 4.5).

At the micro-level analysis it was possible to identify a formal approach to the linkage between business and R & D activities through their business development managers. This type of approach will be referred to as 'personnel'. Conversely, it was not possible to identify any other managerial tool or technique specific to the analysis of such linkage.

Analysis of the 'five competitive forces' (bargaining power of suppliers, bargaining power of buyers, threat of new entrants, threat of substitutes and rivalry within the industry) around Case A provides interesting conclusions, particularly when utilizing the models presented in Figures 3.2, 3.3 and 3.4. In terms of its present main raw material (propylene), Case A doesn't have any major threats as it is supplied by its own corporation (Oilcomp). The only possibility of having major problems with suppliers could occur in new investments (such as in the Far East), in which case suppliers could be considered a threat to Case A's growth. In Figure 3.2 such a 'threat' would be classified within number two (no alternative raw material). The new propane technology will certainly reduce such prospects since Case A will have flexibility to choose its raw material. Such a counterpart fits exactly in group A (modify or develop process to utilize alternative raw material).

In terms of buyers there are also no major threats. However, 60 per cent of the world acrylonitrile output is consumed in the production of acrylic fibres. As shown in Chapter 6, most of the acrylic-fibre manufacturers are presently losing money. Among the threats listed at the right side of Figure 3.2 certainly number five (low profits) reflects the present situation. The new technology will represent a great advantage as it will improve Case A's margins (counterpart C) and consequently enable it to deal with much greater freedom and as a consequence possibly contributing to the survival of the sector.

The analysis of threats and opportunities in the 'new participants dimension' (Figure 3.3) is again highly favourable to the new process. Any new entrant in the acrylonitrile market necessarily has to license Case A's technology. The fact that all world production is based upon this process provides the strongest support for this assertion. The entry barriers are high because of the high level of the technology, patents and legislation protecting the licenser (numbers seven, eight and nine at the top of Figure 3.3). The new process will not only lift such barriers even higher but also lower Case A's exit barriers owing to the availability of alternative raw materials (as described in counterpart B on the list of exit barriers counterparts shown in Figure 3.3). Finally, in terms of threats from substitutes, it is clear that the potentially lower costs of the new process can improve the competitiveness of acrylonitrile derivatives (the discussion about hexamethylenediamine and nylon 6/6 presented in Chapter 7 is a good example of this threat).

The use of the model pictorially represented in Figure 3.4 is again in conformity with the decision taken by Oilcomp and Case A to invest in the new process. The propylene process, although the most modern available, is now more than 30 years old. The diversion of funds towards the new technology fits exactly with what is prescribed on item three of the 'maturity of technology column'. In terms of 'portfolio of activities' the use of any of the matrices there represented would lead to investment and emphasis on the project as the business strength. Additionally, the technical opportunities for the project are high and the industry attractiveness and market opportunity are medium (see McKinsey's matrices within Figure 3.4). Last, looking for what Wilkinson (1987) called 'breadth of activities', it can be seen that the present position of the business would be enhanced

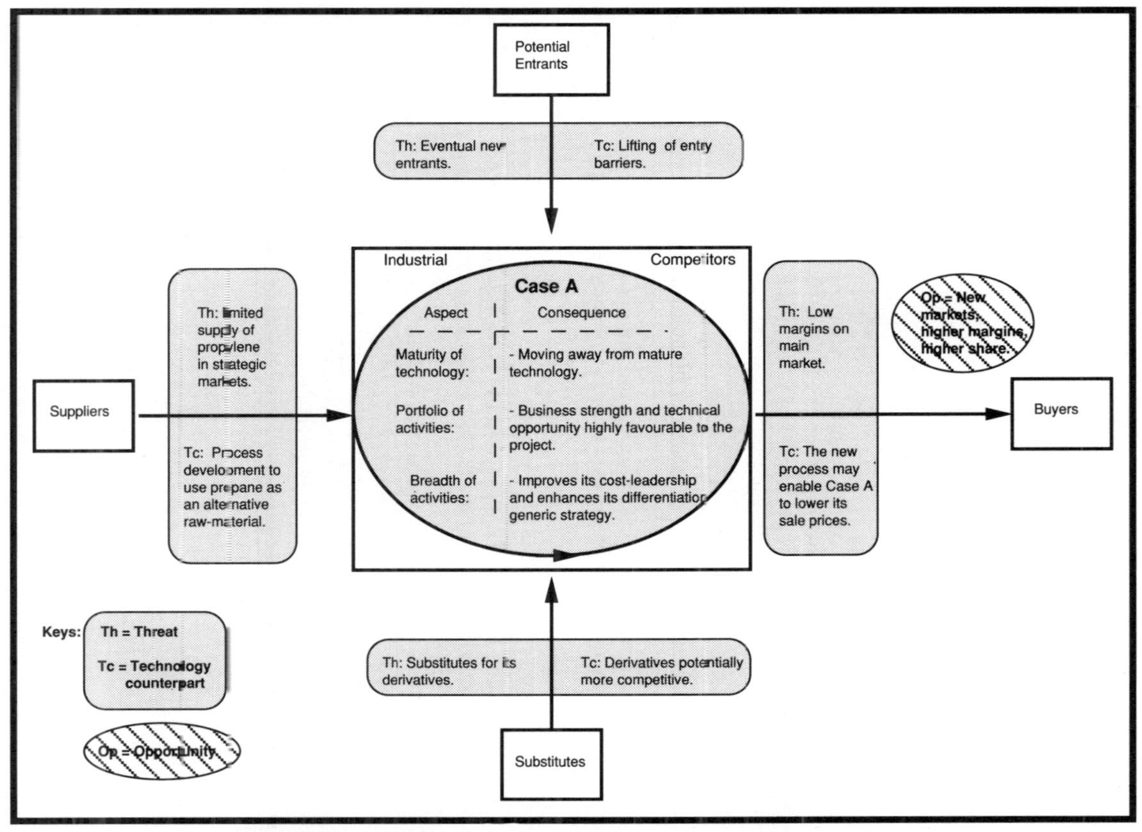

Figure 5.9   Technology and the five competitive forces (Case A).

by the even sharper cost-leadership and differentiation which could result from the new process as already discussed above.

These comments about the relationship between technology and the competitive forces are illustrated in Figure 5.9.

Notes

1. This definition was extracted from page 2 of the annual report and accounts (1988). Similar expressions with the same meaning can be found in 1987 and 1989 issues. The actions related to this strategy, however, can be seen in several earlier annual reports such as 1986 and 1985. At that time such a strategy was not so explicit in the 'Letter from the Chairman', but it was becoming clear through their investments and divestments.
2. Recently, Oilcomp made public its view that 'Nutrition' is no longer a core business for the company. Such a communication was widely understood to be an 'advertisement' that Oilcomp intends to divest that division. Such an intention would appear to be another step in their concentration strategy.
3. The development of analytical methods is a very important part of the R & D activity in the chemical industry but is almost completely ignored in the literature. Basically it consists of the search for and the statistical refining of techniques for qualification and, mainly, quantification of substances or properties present in samples.
4. The forecasting was made by a major consultancy company in the area.
5. This combination of cost-leadership and differentiation strategies was analysed by Porter (1985) as a consequence of technological leadership associated with large market shares fitting perfectly with what has been observed in this case study.
6. An average time lag of three years should be considered between an invention and its respective patent issuing.
7. RM – Reich Mark
8. Figure 5.8 shows some other regular innovations in acrylonitrile catalysts. It also shows a new process for ammoxidation of propane. As there isn't yet any industrial plant using this process, it is not possible to know if it is going to be a revolutionary or architectural innovation, although it can be said that it will eventually disrupt the existing competence.
9. This effectiveness is linked to the realization of the strategy (as shown in Section 1.4.3). The assessment of its efficiency, however, is outside of the scope of this research.

6 Case-study B (acrylic fibres producer)

6.1 Introduction

Case B is a strategic business unit (SBU) within the Fibres and Films division of a large speciality materials company, which will be referred to as 'Fibrecomp'. Within Fibrecomp's matrix framework, Case B has total responsibility over personnel, production, sales and product development of acrylic fibres. Other functions, such as purchasing and research, are shared at division level, and more strategic decisions such as major investment and strategic research are taken at corporate level.

Presently, Fibrecomp has five divisions or business areas: Coatings, Performance Materials, Packing, Chemicals, and Fibres and Films. During the course of this research, however, Fibrecomp demerged its largest division (Textiles, which represented 39 per cent of its turnover and 23,000 employees). This strategic action is important for the understanding of the corporate strategy as discussed below.

Among the five divisions (or business areas) Fibrecomp has 22 SBUs which report directly to the board. Of these, five are within the Fibres and Films division.[1] In Figure 6.1 the outline of the Fibres and Films division is shown within a simplified view of the whole company.

Figure 6.1 is not intended to show a comprehensive outline of Fibrecomp but rather to locate Case B and research and technology- (R & T) related units in the whole company.

At the beginning of this research, Case B had the largest capacity for acrylic-fibre production in Europe. After shutting-down its French plant, however, it lost its position to the Italian group Enichem/Montefibre. In spite of this reduction in capacity (presently 150,000 tonnes/year), Case B maintains its position as the largest merchant buyer of acrylonitrile, since the Italians have a considerable captive production of this raw material (around 150,000 tonnes/year; see Table 6.1). The reasons for the closure of the French unit are linked mainly to the European overcapacity, which is depicted in Section 6.4.

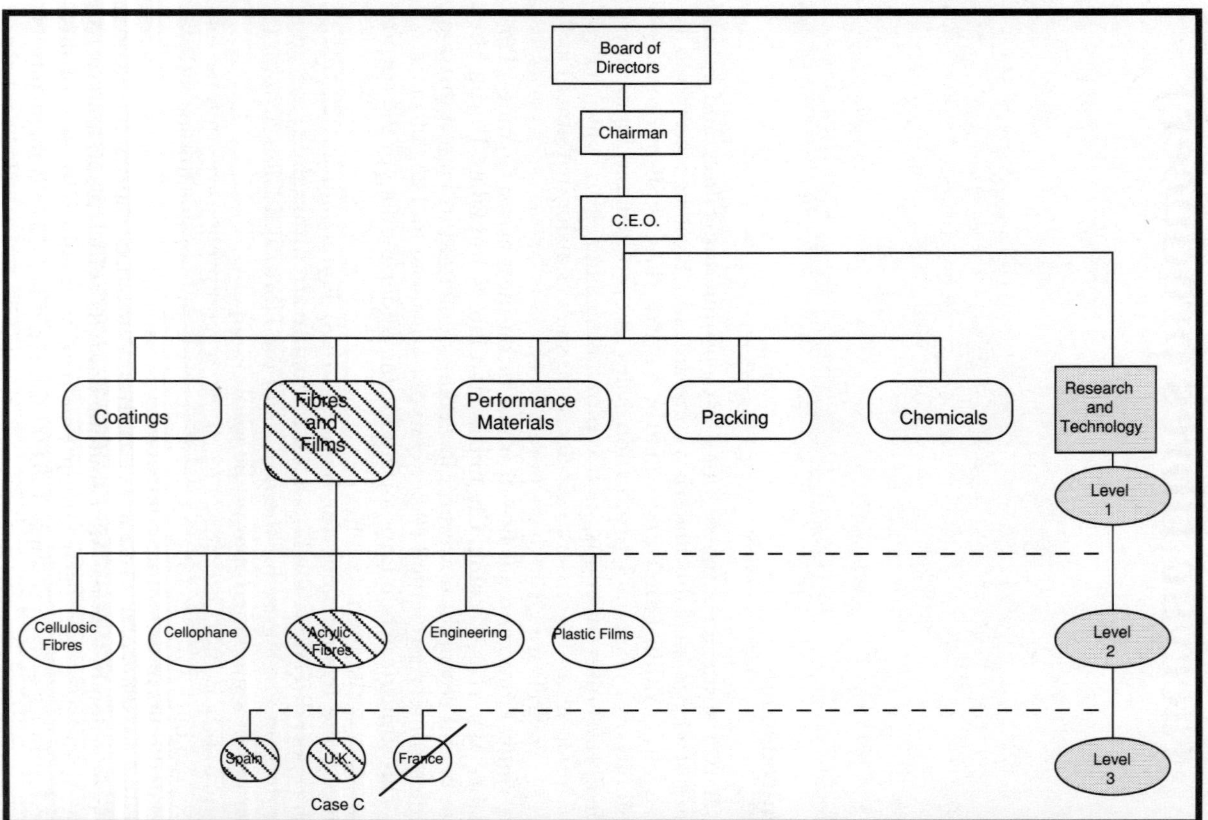

Figure 6.1   Outline of Acrylic Fibres and 'R & T'-related units in Fibrecomp.

6.2 'Fibrecomp's' strategy analysis

Fibrecomp was founded at the beginning of the last century. Its primary activity was silk weaving, and it had a huge growth mainly involving the production of fabrics. The company had a second boom at the beginning of the 20th century, inventing and pioneering the man-made fibre industry. Subsequently, chemistry became a central theme in the company. The activities involved with fibre production overshadowed those of fabrics.

Although heavily dependent upon the British economy, Fibrecomp became an international group as a consequence of the British economic decline from the early 1960s. To defend its customer base for fibre sales the firm diversified, increasing dramatically its activities in textiles.

The 'oil-shocks' in the 1970s heavily affected the European fibre industry, and Fibrecomp was not an exception. During the changes which occurred in the British economy in the 1980s, the need to modernize the management of the company became clear. Major strategic movements began in 1985 with the creation of the 'Textile Group'. This new group brought together dozens of businesses involved in spun yarns, fabrics, home furnishing, clothing and distribution, forming the largest operation of this kind in Europe.

The formation of the Textile Group made clearer the enormous differences between the two types of business (see Lewis, 1989) which had for so long lived painfully together in the firm. While the businesses in the 'Chemical and Industrial Products Group' were capital-intensive, cost-driven and globally distributed, the textile businesses were labour-intensive, fashion-driven and United Kingdom based. At the beginning of 1990, the two groups were demerged, resulting in two completely distinct and independent publicly quoted companies: Fibrecomp Plc and Fibrecomp Textiles Plc. While the 'old' company was almost an industrial conglomerate, the new Fibrecomp can be characterized as an international speciality materials company.

In the 1991 annual report, Fibre Company's chairman stated: 'We are investing for organic growth, reducing our exposure to *adverse cyclical effects in capital intensive industries* and accentuating wherever possible our ability to compete on non-price factors' (author's emphasis). Such strategy seems consistent with the broad concept of a speciality-materials company and with Fibrecomp's further major investments and divestments, as follows:

- Acquisition of 'M Processing Inc', a leading producer of solar films and industrial special films, 1987;
- Acquisition of 'P Paints Company', 'a premium supplier specialised in protective coatings and, architectural paints for the professional market' (*Annual Report 1987/1988*, p. 16), 1987;
- Sale of 'Wood Pulp Business', 1989;[2]
- Acquisition of 'PR Chemical Corporation', an important producer of performance sealants and adhesives, 1989 (see Byrne, 1989);
- Demerging of 'Fibrecomp Textiles',1989.[3]

The acquisitions and divestments listed above were the largest but not the only ones (according to Morris (1989), Fibrecomp spent US$564 million on acquisitions during the previous two years; in the same period they realized US$376 million in the sale of assets). All these investments were made in specialities and the only large sale (wood pulp) a commodity producer. 'Textiles', in spite of the fact of not being a commodity business, was of a very different nature, as discussed above. Evidence of the difficulty of managing such diverse businesses arose a few years ago when, during a meeting of their board, 'only one of the six directors was able to have an opinion about a proposed investment in the textile business', as revealed by an interviewee.

As a consequence of all these strategic movements, Fibrecomp sales in the United Kingdom were reduced from 60 per cent in 1984 to less than 25 per cent in 1990. During the same period, its number of employees fell from 70,000 to 24,000 in 22 businesses, of which at least 16 are involved with specialities of diverse natures but, in principle, linked through core technological skills in polymers, materials, colour and surface sciences. Such changes can be considered consequences of an effectively implemented strategy.[4]

6.3 'Fibrecomp's' R & D structure and strategy

Fibrecomp's global R & D expenditure fluctuated in the last three years of our study, rising from 1.2 per cent of its turnover in 1989 to 2.1 per cent in 1990 and dropping back to 1.6 per cent in 1991 (*Annual Reports*' balance-sheets data). A similar fluctuation occurred in the number of its R & D staff, presently around 750. In both cases, this variance appears to be linked to intense changes in the business portfolio.[5]

The R & D staff, budget and projects portfolio are formally divided into three different levels:[6]

– Level 3: short-term, exclusively business-orientated and supported projects, mainly involving product development through existing scientific knowledge.
– Level 2: Short- to medium-term projects, supported by business areas (such as Fibres and Films for example). These usually involve the application of new knowledge or technologies to existing products in each area.
– Level 1: Long-term projects, involving the development of new scientific knowledge and, in principle, independent of existing products. Nevertheless, these projects are linked to the core technologies in the company (polymer, materials, surface and colour sciences).

Most of Fibrecomp's research (corporate, as well as levels 2 and 3) is located at the research centre in the United Kingdom, although some research is also conducted in the United States, mainly in the 'P- Paint Chemical Corporation' business acquired in 1989. Product development is spread across different sites.

Approximately 15–20 per cent of Fibrecomp's researchers have a doctorate degree, most of them working on level-1 projects. Thirty to 35 per cent

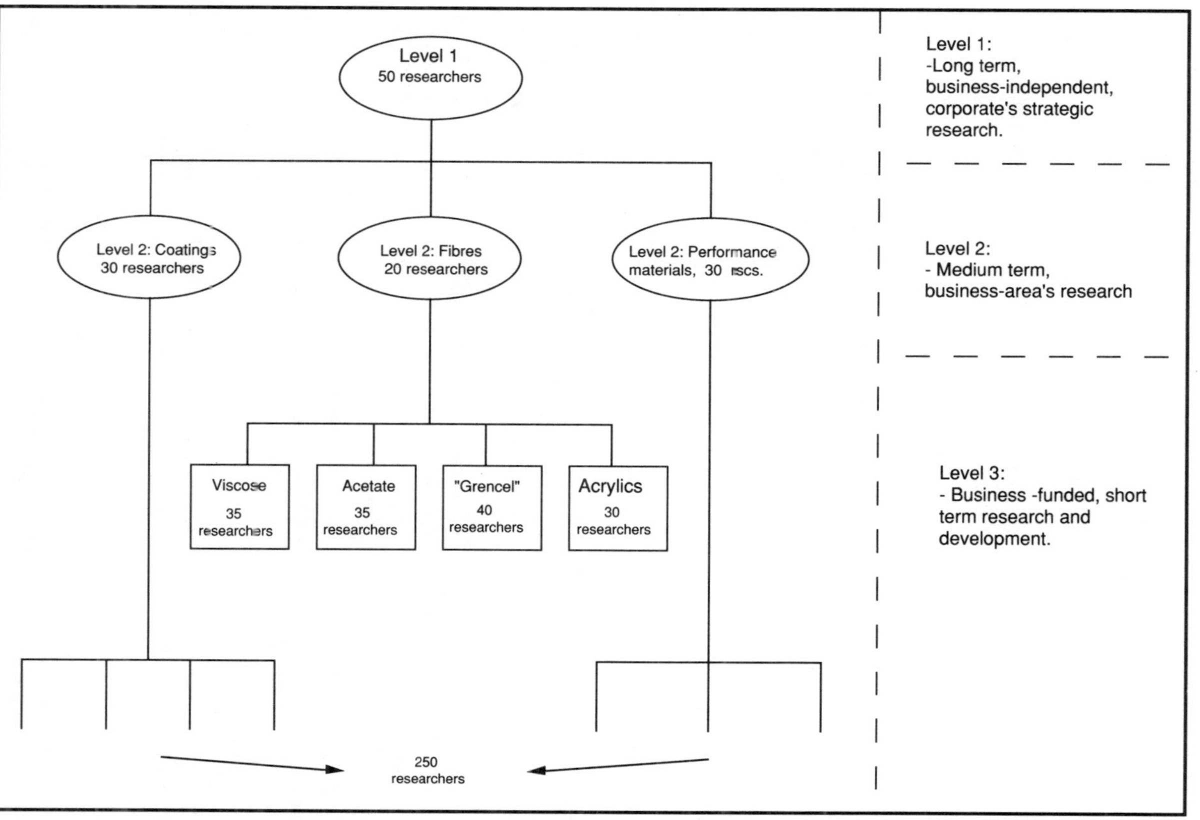

Figure 6.2 Approximate number of researchers per area at Fibrecomp's main research centre.

have a bachelor's or equivalent degree, while the remainder are technicians, administrative and support personnel. At Level 3, Fibrecomp has some 250 researchers; a lower number is shared among the different Level-2 research labels (see Figure 6.2), with some other fifty researchers at level 1.

6.4 Acrylic fibres market review

As discussed in Chapter 3, this research involves products (derivatives of acrylonitrile) in diverse sectors of the chemical industry. Among these products, however, two come under the fibres' label: polyacrylonitrile (acrylic fibres) and polyamide (nylon fibres). As both are within the same sector, it is useful to start their analysis from a common point of view.

For this reason the generic marketing review of both is placed together in this section (Chapter 7 has a smaller, but specific, review on nylon fibres). This review begins with an historic perspective of the world scenario[7] and will lead towards the European and British situation (when and as relevant to the case-study) and will narrow to each particular fibre scenario.

6.4.1 Man-made and synthetic-fibres overview

Fibres of commercial interest can be generically divided into two broad categories: natural fibres and man-made fibres. Man-made fibres can be subdivided again into two other categories: artificial fibres and synthetic fibres.

Within the natural-fibres classification, some of the main examples (for industrial use) are: jute, wool, cotton and silk. The man-made fibres classification, on the other hand, involves a much larger number of different products which will be discussed below.

6.4.1.1 Man-made fibres

All man-made fibres are produced by extrusion techniques and, in principle, any one of literally hundreds of different polymeric raw materials can be used. Owing to the enormous amount of capital and technology involved in perfecting each type of fibre, however, the industry is concentrated around only some of them.

When the polymer to be extruded is based upon a natural polymer artificially modified, the fibre can be generically classified as artificial fibre. Some of the main artificial fibres are viscose, acetate and triacetate.

Viscose was the first industrially successful man-made fibre. Its production process was invented in 1892, but its spinning technique was developed some time later, and industrial production began in 1904–5. The first versions of viscose were known as 'artificial silk', and later as 'rayon'. Today, viscose is available in several advanced forms. It is used alone or, frequently, in blends with polyester or cotton for fashion apparel and workwear, furnishing, home textiles and industrial applications.

Acetate and triacetate are based on a derivative of cellulose (as well as viscose); nevertheless their properties are different. Acetate textiles are silk-like, with soft handle and drape; they are used in outerwear and underwear, velvet-like pile and other furnishing fabrics. Triacetate is used in similar applications, but because of its quick-drying properties is well suited for pleated apparel.

Acrylics and nylons fall into the category of synthetic fibres which is presented below.

6.4.1.2 Synthetic fibres

The history of synthetic fibres began at the end of the 1920s, when Du Pont's chemical department received the approval from its company's board of directors for a fundamental research programme, related to 'rubber-like substances', to be developed by Dr Wallace Carothers. The main output of this programme was the first fully synthetic fibre, which was called 'nylon'. Full industrial-scale production of synthetic fibres began in 1939 with Du Pont's nylon 66, followed in the same year by I.G. Farben's nylon 6 (Perlon).

The 'revolution' triggered by nylon fibres led to ICI's development of polyester fibre during the Second World War and its commercial production from 1949. In 1948 Du Pont initiated its acrylic fibre (Orlon) production, immediately followed by the I.G. Farben descendent Bayer with its own acrylic fibre (Dralon).

Today there are several different synthetic fibres in the market-place of which some of the more important (from a commercial point of view) are polyester, polyamide (nylons 6, 66, etc.) and acrylics.[8] Synthetic fibre's biggest market is textiles. From this market point of view, the main characteristics and applications of these fibres are described below.

Polyester is currently the most widely used synthetic (and man-made) fibre, accounting for almost 20 per cent of all fibre used in the world. As with almost all modern fibres, it is available in many versions but all with a common strength and sunlight resistance. Applications of polyester range from curtain fabric to conveyor belting.

Polyamide fibres have a higher number of different versions than any other fibre. Even with the basic molecular structure of the polymer used in their production, there is a great variety (such as nylons 6, 6/6, 6/9, 6/10, 6/12). Polyamide fibres can be strong yet comfortable for textile end-uses. They resist bacteria and chemicals, becoming one of the most popular fibres in carpets, underwear, lingerie, swimwear, hosiery and sportswear. Industrially, they have demanding applications such as tyres, ropes and tenting fabrics.

After polyester and polyamide, acrylic fibres are the third most important group. They are manufactured fibres in which the fibre-forming sub-

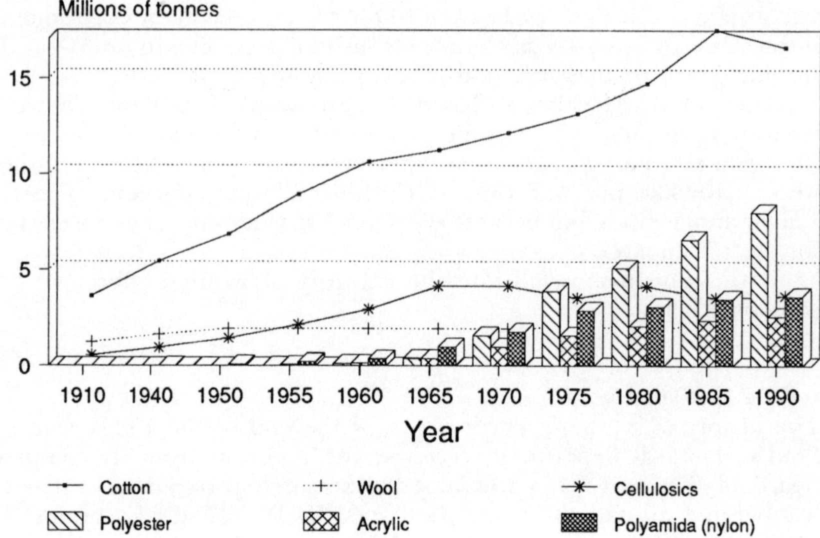

Figure 6.3 World fibre consumption (author's estimate from several sources).

stance is any long-chain synthetic polymer composed of at least of 85 per cent (wt per cent) acrylonitrile ($CH2=CHCN$) units. When the percentage of such units decreases to a range between 35 per cent to 85 per cent , they are defined as modacrylic fibres (in this book both modacrylic and acrylic fibres are generally classified as 'acrylic fibres'). Their main characteristics are linked with their wool-likeness. Their comfort and warmth are attractive in knitwear, their ease and resistant dyeing make them very popular in upholstery, velvets and curtains. Acrylic fibres are also widely used for hand-knitting yarns. Garments possess easy-care qualities and maintain appearance over time. Modacrylic fibres are inherently flame retardant and are used for furnishings, sleepwear, toys and fake fur.

6.4.2 Fibre industry: industrial linkages and market consolidation

The fibre industry is an extremely complex sector of the economy. It is restricted by, and intimately linked with, its main 'up-stream' and 'down-stream' industries: chemicals and textiles. There is even a third industry on which it depends, fibre and textile machinery. Technologically speaking, the fibre industry is on the edge of the chemical-process industries, but strategic moves in any of the other two segments usually have important consequences for it, as will be shown below.

In spite of their qualities, man-made fibres did not gain a major share of the market until the 1960s. The main reason for such slow growth in the man-made and, mainly, in the synthetic fibres market was a 'technological bottleneck' represented by fibre and clothing machinery. Until the 1950s such machinery was designed for natural-fibres processing and, at best, was only adapted for synthetics. Such strategic dependence upon the machinery sector posed a threat to the development of this new industry. However, from the 1950s, the increasing popularity of man-made fibres encouraged machinery builders to introduce appropriate equipment, exclusively designed for synthetic fibres production. The uniformity of the fibres facilitated the development of improved processing systems. These improvements allied to increased recognition of the qualities and potential of these fibres have increased the rate of international use of man-made fibres.

As can be observed from Figure 6.3, in the first decade of this century total output usage of fibres consisted almost entirely of wool and cotton. During the following four decades, cellulosic fibres, such as viscose, made considerable progress. In the mid-1950s, synthetic fibres started to make an impact. During the next 30 years, man-made fibres (mainly synthetic) formed half the total consumption. From 1980 to 1990, however, this growth was much more dependent on synthetic's performance over cellulosic (or 'artificial fibres'), with almost no change in the ratio for natural/man-made fibres' world consumption.

6.4.3 The different stages of market development

As cited above, the world-wide ratio of natural to man-made fibres production remained almost unchanged over ten years. The ratios in 1980, 1988 and 1990 were respectively: 1.09, 1.09 and 1.1. While global data has been relatively constant, the ratio cited has been changing quite dramatically in the different parts of the world. This has enormous implications for the industry, as will be shown below.

In general, it can be said that the fibre market has gone through three different stages from the 1950s until the present. The first period could be called 'growth' and was characterized by a strong effort by each producer towards generic growth. During this period, each producer manufactured different fibres and strived to sell each of them to all types of markets. After the first oil shock (1973), the industry moved towards what can be called its 'rationalization' period, during which considerable effort was aimed at a rationalization of production capacities and scopes.

In Europe, the EEC took the initiative, through the International Rayon and Synthetic Fibres Committee (CIRFS), which had representatives from all the major producers. As a result, European production suffered a 12 per cent reduction in capacity (400,000 tonnes) during 1978–9. At about the same time, the volume of imports from 'low-cost countries' was regulated by the first Multi-Fibre Arrangement (MFA) signed in 1974.[9]

During the 'growth' period (ending in 1973), capacity utilization in European plants was around 92 per cent. After the first oil shock, capacity utilization dropped to 76 per cent, falling even further to 61 per cent in 1976 (Davies, 1991). In such a capital-intensive sector, this low level of utilization can have serious consequences. Davies estimates that in 1975 the European fibre industry lost US\$720 million as a consequence of its overcapacity and the high costs of raw materials (petrochemicals).

Table 6.1 Capacities of the major West European man-made fibre producers in thousands of tonnes, 1990 (adapted from Davies, 1991).

	Acrylic	Polyester	Polyamide	Cellulosic
Enichem/Montefibre	350	165	–	–
Hoechst	70	230	5	65
Rhone-Poulenc	–	130	140	45
Akzo	–	185	110	100
'Chemicomp'	–	20	200	–
Lenzig	–	–	–	130
Kemira	–	–	–	70
Svenska Rayon	–	–	–	30
Snia	–	–	90	10
'Fibrecomp'	150	–	–	70
Bayer	150	–	–	–
Total	720	730	545	520

A second agreement between the major European producers (October 1982) led to another 500,000 tonnes' reduction in capacity. This second agreement appears to mark the beginning of a third stage, which can be termed as 'specialization'. The second agreement was qualitatively different from the first, once the capacities were reduced towards specialization. Each producer cut capacity in just one or two types of fibre, opening the way towards specialization for each company. During the following years, several businesses were bought and sold among major European producers. As a consequence, each producer presently focuses on just one or two synthetic fibres, as shown in Table 6.1.

The third stage meant that companies not only concentrated on only one or two fibres but also that there was an increase in the specialization of end-use for the different fibres.

– Nylon (polyamide), for example, seems to be targeted at hosiery, furnishing and industrial yarn – applications in which its characteristics of lightness, bulk stretch recovery and durability provide excellence.
– Polyester is the general-purpose fibre and a 'natural partner' for blends with cotton, taking care of regular textile needs.
– Polypropylene is a 'new entrant', at present used mainly for non-wovens and where cover at a price is the determinant.
– Acrylics with its wool-like properties is mainly used for knitwear, jersey and finishing fabrics (partly owing to the fire resistance of modacrylics).

6.4.4 Cotton's 'turnaround' and the competition from developing countries

At present there are two main competitive pressures influencing the synthetic fibres' market. These are: the growing threat from 'low-cost countries' and the turnaround by cotton fibres.

From the beginning of the 1970s to the 1990s Japanese production of man-made fibres grew at an average rate of 1.2 per cent per year, compared

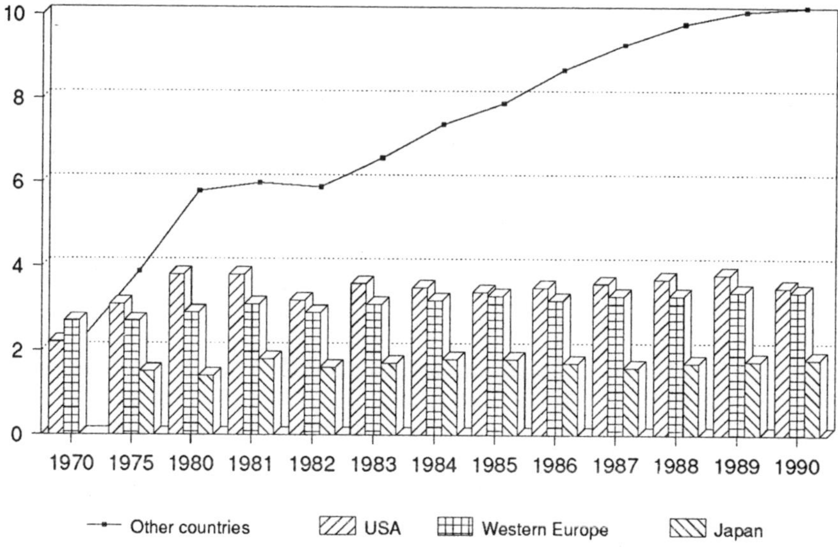

Figure 6.4 World production of man-made fibres, 1970–90 (adapted from Davies, 1991).

with 1.8 per cent per year in Western Europe. During the same period, American production grew at an average rate of 3.2 per cent , while global production by all other countries grew at the staggering average rate of about 20 per cent a year, meaning an overall growth of 500 per cent in twenty years. This dramatic change in the world scenario is illustrated in Figure 6.4. The reasons for such a change seem to be linked to two main factors: the availability of technology and the great impact of increasingly favourable exchange rates of developed countries' currencies against those of the developing world.

The availability of technology seems to have begun in the 1970s. Previously, the technology for man-made and synthetic fibres remained with established producers. During the 1970s, many engineering companies acquired and developed sufficient modern technology to be able to offer entire plants with state-of-the-art equipment. The recession caused by the oil crisis first affected the developed countries (at that time already having an overcapacity in fibres), dramatically cutting the engineering companies' market. At the same time, the economic model of 'imports substitution' was spreading throughout the more dynamic economies of the Third World, creating a new and large market for contractors.

During the same period the West European currencies, together with the Japanese yen, experienced a great increase in value relative to those of developing countries. At the end of that decade the American dollar also began a period of great growth, retaining the conditions for a continuous expansion of exports from developing to developed countries for almost 20 continuous years. The effect of currency exchange rates upon the competi-

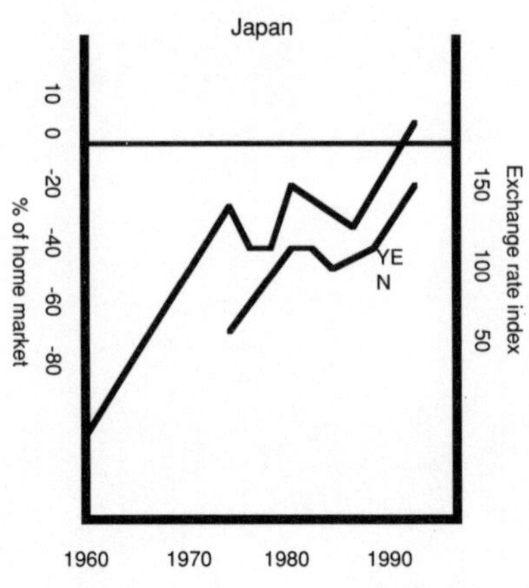

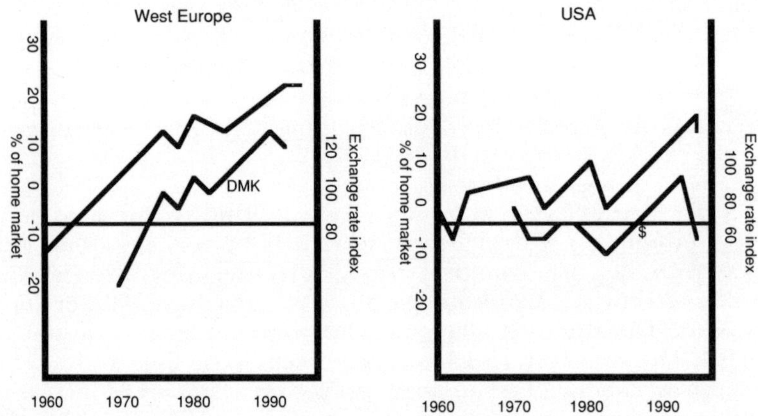

Figure 6.5   Imports (net of exports) as a percentage of home market and exchange-rate change (adapted from Buck, 1989).

tiveness of fibre producers was demonstrated by Buck (1989) and is illustrated in Figure 6.5.

An additional competitive threat came from changes of end-user behaviour in the developed countries. While synthetic fibres, in general, have better physical and chemical properties, cotton seems to be recognized as being more comfortable, especially when in direct contact with skin. This situation seems to produce some type of equilibrium between the demands for natural and man-made fibres. One historically important factor in the balance between them, however, is their relative prices. As shown in Figure 6.6, consumption in developed countries has moved away from man-made fibres since the second half of the 1970s (Japan) and the first half of the 1980s (mainly Western Europe and The United States). At the same time, developing countries are continually moving towards increased use of man-made fibres. It seems that people in developed countries 'can afford wool' and, as discussed by Buck (1990), are changing their taste to the aesthetics

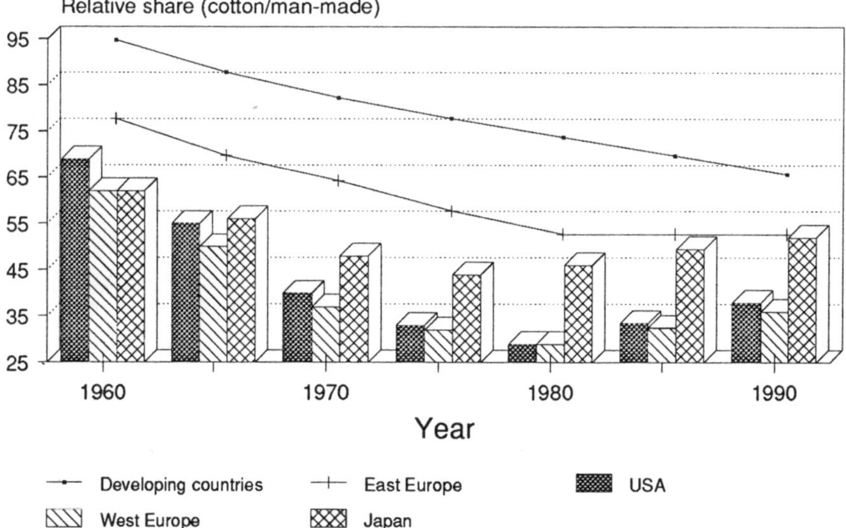

Figure 6.6   Relative share of cotton and man-made fibres in different economic regions (adapted from Buck, 1989).

(and comfort) of cotton (and, in Japan, to a certain degree towards silk) even if it is less practical to wear, less durable and more expensive. A second factor may be the identification of cotton fibres as 'greener' or more environmentally friendly.

6.4.5 Acrylic fibres

The main characteristics of acrylic fibres (from a commercial viewpoint) are their likeness to wool, softness, launderability, shape retention and dyeability, which enable these fibres to be recognized in the market-place as an 'affordable wool'. As a consequence, its main markets are pullovers and fleecewear. In Europe, for example, acrylics have almost 70 per cent of the pullover market. Data relating to European consumption of fibres by end-use show market share as generally constant since 1980, characterizing a mature market. Apparently, the optimum fibre for each end-use has now been largely determined and the process of substitution of one fibre for another completed. Figure 6.7 shows the distribution of acrylic by major markets (apparel, home furnishing and industrial) in Europe and in the United States. It is clear that the distribution between these markets is very similar, showing a tendency to equilibrium at around 73 per cent in apparel. Considering that pullovers represent almost 25 per cent of American and 32 per cent of the European consumption of acrylic fibres, it is easy to see how dependent acrylic fibres are upon fashionable goods. During 1987, for example, all man-made fibres, with the exception of acrylics, were in

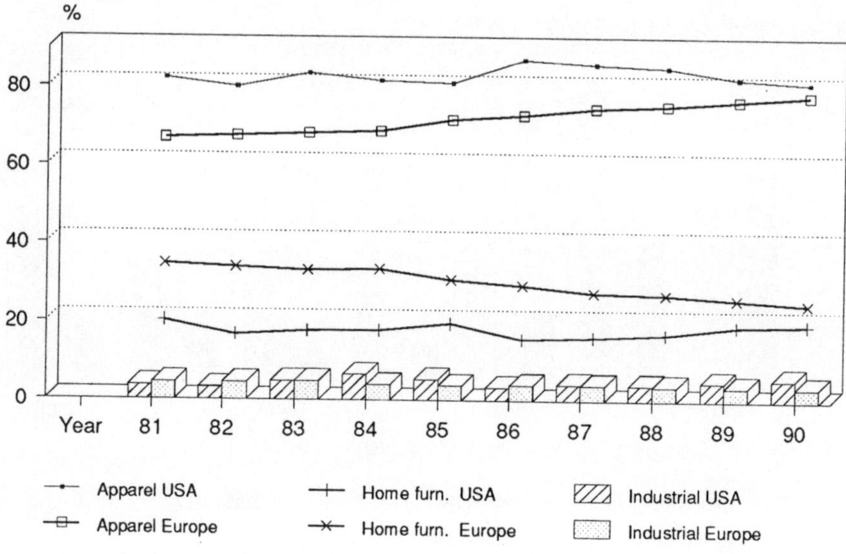

Figure 6.7   Consumption of acrylic fibres by major markets in the United States and Europe.

high demand. The poor demand for acrylics was attributed to the fashion move away from 'chunky' pullovers. Another good example can be seen in Figure 6.8, which shows the strong growth and decline of the fleecewear share of acrylic-fibre consumption in the United States (from 1984 to 1989; it rose from 57,000 to 93,000 tonnes, dropping back to 35,000 tonnes). Another important issue of fashion is related to colour. Some 95 per cent of all acrylic fibre produced throughout the world is undyed. The other 5 per cent , however, is 'producer dyed' and has a typical price premium of 40 per cent when compared to undyed fibre. To producers competing for this niche of the market, the influences of fashion can be very important, as will be shown in sub-sections 6.6 and 6.7.

While in the long term fashion is not seen as so important, from day-to-day it is quite influential. Adding to the effect of fashion is another significant aspect of the fibre industry: the length of its supply-chain, passing from fibre producers to the textile industry, clothing manufacturers, distributors, retailers and finally end-users. This characteristic creates an enormous inertia in terms of response from fibre producers to eventual changes in consumer taste. As a consequence, fibres seem to pass through alternate cycles of four to five good years followed by a similar number of bad years. Although this aspect creates another complicating factor in the management of this business, it creates opportunities for those companies and countries with the capability to have a faster response time and flexibility. The discussion of such opportunities will be resumed in Section 6.7.

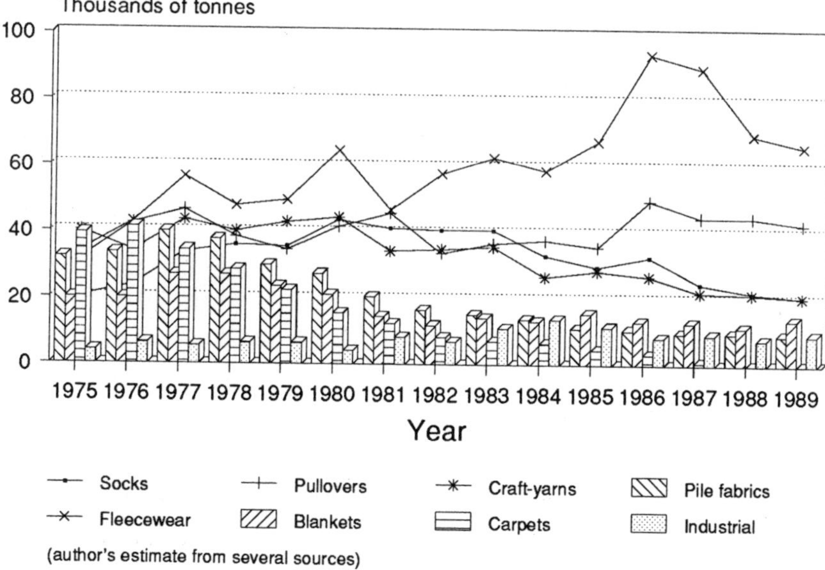

Figure 6.8 Consumption of acrylic fibres by end-use in the United States.

6.5 Acrylic fibres technology review

The manufacture of acrylic fibres necessarily involves two main steps: the manufacture of the polymer and the manufacture of fibre.[10] In the first the basic molecular structure and chemical characteristics are defined. In the second, the filaments are produced from the viscous raw product of the initial reaction. Between these two major operations, but mainly after the second, several steps are necessary in order to produce the enormous variety of effects which will make each fibre appropriate to each specific end-use. Figure 6.9 illustrates, in a simplified view, most of these steps.

Polyacrylonitrile is produced basically from acrylonitrile (83–92 per cent), other acrylates (such as methyl acrylates, 6–8 per cent) and other monomers (3–5 per cent). There are four different processes to conduct this polymerization but only two have major economic importance: suspension and solution polymerizations. In both cases, the first step (after polymerization) is the separation and recycling of unreacted monomers.

There are also two basic types of spinning: wet and dry. The nature of the spinning process defines the operations to be conducted after the spinning. Similarly, the method of polymerization will define the unitary operations to be conducted before spinning. As the solvents used during the solution polymerization[11] may be the same used during spinning, this second type of polymerization (solution) implies a smaller number of operations and, consequently, lower costs. Suspension polymerization, however, allows greater flexibility.

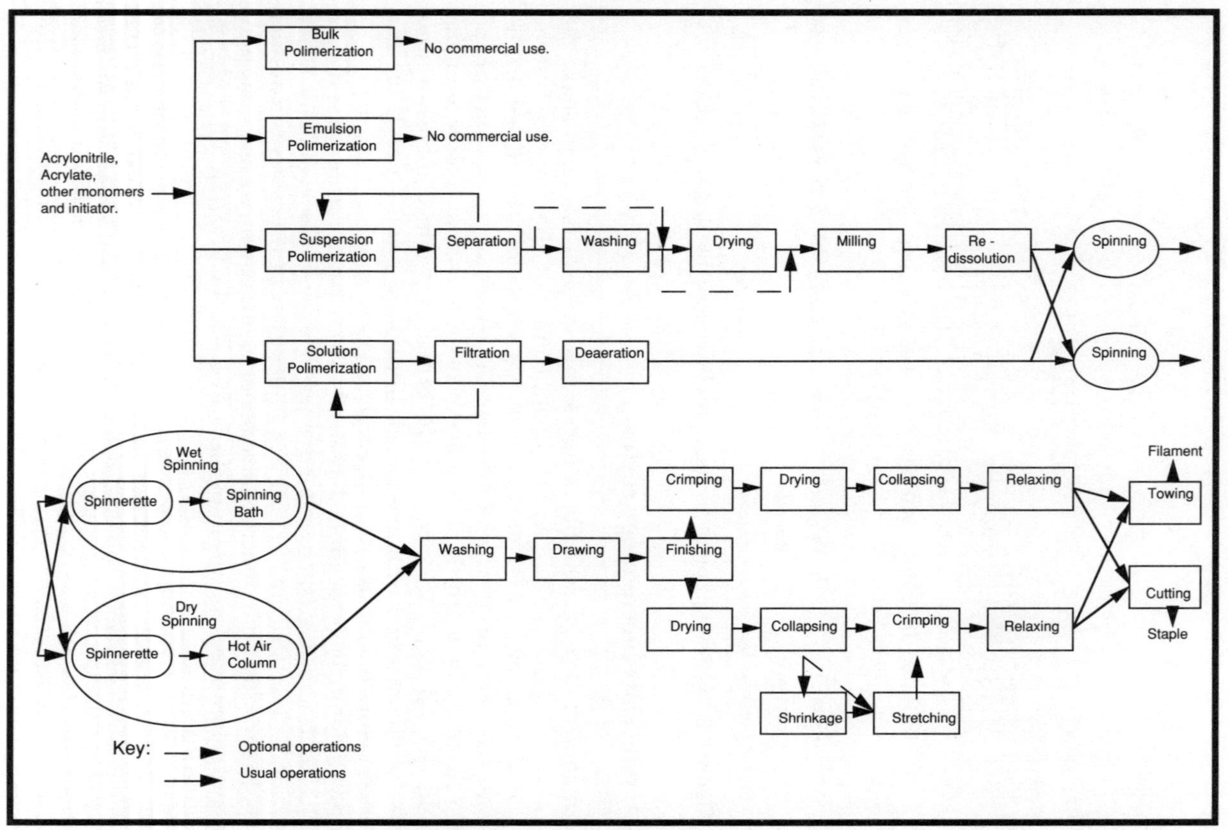

Figure 6.9   Acrylic-fibres manufacturing process.

Even considering only processes with the same type of polymerization and spinning, it is possible to have considerable differences through the process as a whole (as illustrated in Figure 6.9).

Immediately before spinning, numerous additives are included. Such additives can be chosen from a large range of options and for a range of purposes (flame retardants, pigments, delusterers, stabilizers, etc.). Both types of polymerization can, in principle, be used with either of the spinning processes. A large majority of the producers (approximately 80 per cent) use wet spinning.

After spinning, there are three compulsory steps:

Washing – which is the operation responsible for the removal of the spinning solvent, dope and other materials.

Drawing – when the fibre is stretched to extremely controlled tensions in order to create some molecular orientation in the fibre, consequently enhancing its physical properties.

Finishing – when additives are included. Typical finishes include a lubricant to prevent filament-sticking during drying, softeners and anti-static additives. Although these three steps are obligatory, the way in which it is done varies quite dramatically from producer to producer.

After finishing there are still more variations as the sequence of operations can be completely different. The group of operations conducted between finishing and cutting (or towing) is sometimes referred to as 'drawing' but is composed of at least four different physical processes: drying, collapsing, crimping[12] and relaxing (some optional processes can still be added such as shrinking and further stretching). The drying, collapsing and relaxing operations can sometimes be conducted in only one device. After this process the fibre will possess far better elongation and dyeing diffusion ability as well as an enhanced resistance to fibrillation and shrinking.

In the last step, the fibre will be 'towed' or cut as staple (in lengths of 2.5 to 15 centimetres) and packed to be sold.[13] Only after sale, when already on the weaver site, will the fibre be processed to become a yarn or a fabric.

As can be concluded from this overview of fibre processing, the number of combinations is almost infinite. To illustrate such an assertion, one of the managers interviewed during the conduct of this case-study observed that just by combining some of their variables his company could, theoretically, produce more than 36 million different types of fibres!

In all of the descriptions of fibre manufacture above, one operation was intentionally omitted, namely the dyeing process. As described previously, almost all world acrylic fibre production is sold as undyed fibre. When an undyed fibre is bought, the buyer can choose among dyeing the staple (before it is spun into a yarn), dyeing in yarn cones, as a fabric or even finally as a garment. Producer-dyed fibres used to be dyed during the fibre spinning or immediately before collapsing (tow). The dyeing process has an important role to play in the understanding of this case-study and will be discussed in sub-section 6.7.2.

6.6 Case B's business strategy analysis

Case B began its commercial operations in acrylic fibres in 1956 and a little later (1962) made an important innovation in this process which still has an important influence on its strategies. At present, Case B has all its production in England (as cited before, during the course of this research it closed its plant in France).[14] While Case B's main market is in the United Kingdom, it also has an important role in Continental Europe, the United States and China. In all these areas, Case B sells to different segments and types of end-use.

In order to understand Case B's strategic position, it is necessary to be aware of its production process as discussed above. As mentioned in Section 6.5, the colours of a garment or fabric can be defined after or before its production. In the former case it relies on the use of undyed fibres, while the latter is dependent upon dyed ones. In contrast with other acrylic-fibre producers, Case B has more than 75 per cent of its sales based upon dyed fibres which have, for obvious reasons, higher prices. This observation portrays a focus on dyed fibres which, associated with a sales policy of premium prices, characterizes a generic strategy of 'differentiation-focus' (as defined by Porter, 1985). From the fibre buyer's point of view, this premium results from two main factors: quality and time of response, as explained below.

As a result of its processing technology, Case B is the only producer able to dye consistently and price competitively its fibre using an on-line process. This particular technology makes Case B capable of producing smaller 'batches' of specific colours and shades significantly faster (and at relatively lower costs) than its competitors. Such characteristics allow its customers to keep smaller inventories or even to have no inventory at all. The same technology enables Case B's dyed fibres to have better and more consistent colour quality, providing the conditions necessary to keep Case B's differentiation (in spite of its lower production costs, Case B practises a policy of premium prices).

Considering Fibrecomp's aims and generic strategy, Case B is, in principle, a contradiction in the light of its capital intensiveness and the cyclic nature of its market. This contradiction is, however, only relative as Case B's products are also sold based upon non-price factors in line with the corporate strategy as stated by its chairman in 1991 (see Section 6.2).

6.7 Case B's technology strategy analysis

Case B has some 35 researchers working at Level 3, while the Fibre and Films division shares another 20 researchers or so at Level 2 . Owing to its condition of maturity, it is very probable that acrylic fibres do not play a considerable role in Level 2's project portfolio.[15]

Case B's R & D effort is directed towards product as well as process development, as depicted in the different levels of analysis presented below.

6.7.1 Macro-level

As shown in Section 6.4, the main threats to the synthetic-fibre industry are presently the competition from 'greener' and 'better-feel' fibres such as cotton. To European, American and Japanese industries, there is a second main threat – the developing countries' fibre industries. Apparently these threats are enough to justify the present technological trends or 'trajectories' towards specialization and, in some cases, towards smaller fibre diameters (micro-fibres).

While the trend towards specialization can be explained by the market conditions (Section 6.4), the observation about fibre diameters requires further discussion.

In the last three or four years micro-fibres (fibres of less than one denier or one decitex[16]) have been the 'big story' in the industry as referred to by Davies (1991). Such micro-fibres are finer than silk, which is nature's finest fibre. This characteristic opens several possibilities in terms of market, such as:

- 'Water-proof/breathable' fabrics;[17]
- Higher moisture-absorption fabrics;
- Silk-like fabrics.

While the first characteristics are very useful, they are also relatively limited in terms of end-market. The silk-likeness (or at least a better feel and look), however, is desirable and useful for several applications and markets. On the other hand, micro-fibres have a tendency (intrinsic to their small diameter) towards so-called 'pilling': the formation of small 'balls' of fibre due to friction between a fabric and other surfaces. This process of pilling is one of the main problems for any textile company, particularly those producing 'wool-like' products. Notwithstanding fibre thickness, the chemical nature of a fibre is equally important to 'touch' and 'appearance'. While acrylic fibres can have some advantage over other fibres in terms of colour and appearance, the same cannot be said in terms of comfort against the skin. Such a disadvantage, allied to the pilling process, means that micro-fibres are (in principle) a restricted application for acrylic fibres.

Case B apparently recognizes these facts since it is not investing on such a trajectory, although their managers state that they possess the technology for micro-fibres processing. In spite of the fact that Case B is apparently not interested in following such a trajectory, it can still be associated with the broader trends towards specialization.

6.7.2 Meso-level

As described before (Section 3.2.2), the meso-level is the main locus where interaction and competition between different companies and their strategies occur. It is also the analytical space where these strategies can be linked with the opportunities provided from technological trajectories and cultural changes.

As described previously, Case B's technology enables a differentiation-focus strategy. Such technology is, however, nearly 30 years old. This means that it was developed in a period when differentiation was not a common trend but could be a 'natural trajectory' in accordance with Dosi's concept (1982, p.148). According to information collected in the interviews for this case-study, it was already the goal of this process research and development to have a major flexibility in terms of colour change. Independent of past intentions, Case B built a flexible plant and the market moved to the present scenario, where specialization is not only a common trend but also the key to profitability.

The interviews conducted in Fibrecomp's research department indicate that a major part of their effort in the area of acrylic fibres is presently directed at reducing the 'dead time' between the processing of two different colours. (While at the beginning of its utilization the process needed to produce at least 80 tonnes of product per shade, it presently needs about one tonne.) In addition to these projects, they mainly research fibre finishing and colour technology. The latter is also a competitive advantage: while any company can put effort into researching and controlling colour, Case B can go beyond this by using their files with more than 15,000 shade formulations historically accumulated (owing to the peculiarities of their process). In order to understand why these historically accumulated formulations are a real technological asset, it is necessary to discuss briefly the acrylic-fibre dyeing processes.

As described in Section 6.5, there are two basic types of 'producer-dyeing processes'. The first involves the addition of pigments to the spinning solution and results in very good colour retention, although it necessarily implies enormous amounts of fibre per shade (not less than 50 tonnes). This process is used by Bayer in Germany and, until some months ago, by Hoechst (which seems to be steadily withdrawing from acrylic fibres).

The second basic process is the dyeing of fibre as a tow. This process is not only used by Case B but also by Montefibre. In this stage of its manufacture the fibre possess an ideal porosity, allowing a good uniformity on dye diffusion throughout the fibre. While both processes are similar (in terms of the stage when the fibre is dyed and also of being an on-line process), Montefibre's seems to need not less than 10 tonnes of fibre per shade. This is the point were Case B 'colouring culture' makes the difference. As a result of their huge data base of colours and formulations, Case B is capable of recycling the 'head' and the 'tail' of each shade's batch to obtain other shades, allowing much smaller (and consequently flexible) lots. Such a process seems to have other hidden peculiarities which allow, in contrast to other processes, negligible consumption of energy during the dyeing process.

A second point at which their business is intimately connected to technological opportunities is in raw-material purchasing. Some 20 years ago, Case B was in a strategically weak position in relation to a raw-material supplier (Case A above). While Case B's production process makes it strategically strong in terms of product quality and cost, it also made it strategically weak in terms of supply, as one of the process limitations is its sensitivity to acrylonitrile contaminants. In using acrylonitrile with a level higher than 'X' ppms (parts per million) of an individual contaminant, their fibre will present distortion in their colours and shades. Such a limitation made Case B, at that time, dependent upon supply from the only producer with acrylonitrile lower than 'X' ppms of that contaminant's concentration (Case A). As a consequence, Case B initiated a research project which resulted in its own process and equipment for 'extra-purifying' acrylonitrile (as discussed in Section 5.6). This technology presently enables Case B to buy cheaper acrylonitrile on the 'spot market', unlike other fibre producers which rely upon higher contractual prices for their supply. According to data collected during the interviews, Case B saves several million pounds Sterling each year as a result of this policy, whose analysis will be resumed in Section 6.8.

6.7.3 Micro-level

None of the techniques, tools or decisions-aids introduced in Section 3.2.3 are in use at either Fibrecomp or Case B itself. Neither have a department such as Case A's business development (which, in that case, has the formal responsibility to promote the interaction between the R & D and business functions). Fibrecomp's own research organizational structure, however, seems to be a strong factor in promoting such interaction:

- The short-term projects (mainly product and process development) are conducted by each business's own personnel within the corporation's research department (Level 3).
- These individual business projects are connected and complemented by each business area's (Level 2) research. For Case B, for example, it is fibre technologies in general (several process operations are common such as spinning, dyeing, creeping etc), favouring an intense synergy between the different businesses within the same area.
- At Level 1, the long-term research is conducted independently of any individual business or even business areas; nevertheless, this 'basic' research is conducted within the core sciences of all of Fibrecomp's businesses – polymers, materials, surface and colour.

6.8 Conclusion

The movement of Fibrecomp corporate strategy towards specialities and internationalization is now advanced and apparently near its completion. They considerably reduced their 'exposure to adverse cyclical effects in capital intensive industries' as stated by the chairman in the 1991 *annual report*. The new business portfolio is considerably smaller than five years before and certainly more homogeneous. Such a strategy seems to have been more deliberate than that of Case A, as some profitable businesses were divested in order to achieve this homogeneity and all businesses are now linked through a consistent core of technologies involving polymers, materials, surface and colour sciences. This impression is reinforced by the drastic decision of demerging its textile businesses.

Fibrecomp's corporate R & D expenditure is considerably lower than average in the speciality chemicals sector (see Olin, 1972, and Leadbeater, 1991). It is, however, well aligned with the corporate strategy for what its organizational structure certainly contributes.

In terms of Case B itself, the macro-level technology-strategy analysis presented a natural trajectory towards specialization in the fibre industry as a whole and towards thinner fibres (particularly in non-acrylic segments).

The meso-level technology-strategy analysis showed that this business, in spite of its intrinsic commodity-like economic cycles, is running under (and towards) a differentiation strategy, which has assured profits even when most of its competitors are losing money. Such a performance is a result of several factors (see *Financial Times*, 1991c) but certainly involves an intimate linkage between its technological peculiarities and a business strategy tailored to capitalize on such peculiarities, not only in terms of their dyeing technology (which may be related to the concept of technological traditions) but also in terms of their raw material treatment processes.

The micro-level analysis shows an organizational structure of Fibrecomp's and Case B's R & D which keeps business and technology developments in intimate contact with each other. This structure also seems to promote an adequate balance between short-, medium- and long-term research projects. In principle, such a structure should make the sharing of information easier, increasing the probability of deriving a business strategy from technological advantages and vice versa. The model seems to be able to avoid the mismatch between technology and product life-cycles suggested by Coombs and Richards (1991) and discussed earlier in Section 2.4.2. Such an approach to the linkage between technology and business strategies will be referred to in this book as 'organizational'.

None of the innovations here analysed (acrylonitrile purification and the dyeing process) disrupted either the existing 'market/customer linkages' or 'existing competence', both classified as regular innovations in Abernathy and Clark's (1985) taxonomy. In spite of this classification, these innovations are fundamental to the accomplishment of Case B's strategies.

The analysis of the effects of Case B's technology upon the five competitive forces around it proved quite fruitful. The clearest example of such effects can be taken from the 'on-going flow of products' dimension' (Figure 3.2). The research project which generated its 'ion-exchange treatment' for acrylonitrile purification is clearly a counterpart to the strength of its former supplier and exemplifies the use of the model pictorially represented in Figure 3.2. From a position of strategic weakness, in which it had to buy from one single and powerful supplier (factors 1 and 2 of supplier strength in the model), Case B developed the process to utilize alternative suppliers (counterpart A). The business strategy derived from this technological capability enables Case B not only to avoid buying from

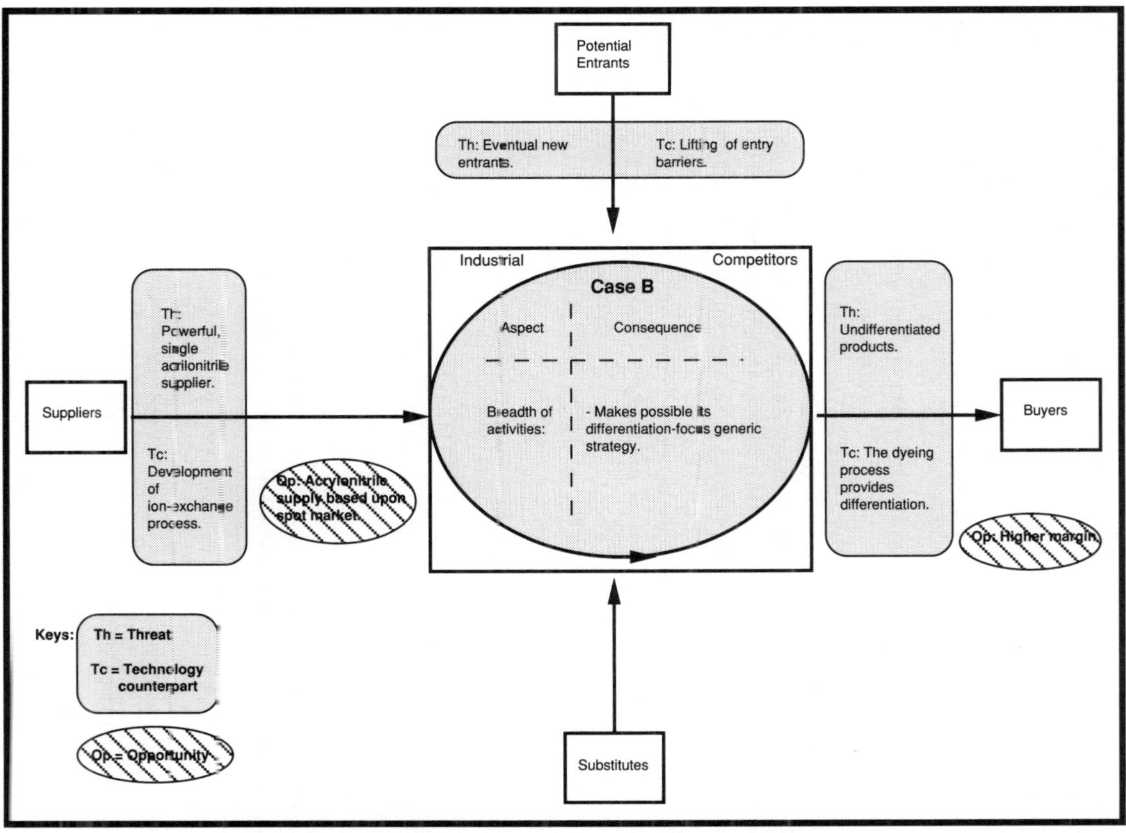

Figure 6.10 Technology and the five competitive forces (Case B).

its former supplier (Case A) but also allows this company to buy massively in the spot market, enhancing its profitability significantly.

In terms of buyers, the dyeing technology enabled Case B to minimize product competition (number 3 in the model) through counterpart B (creating product differentiation). In the 'new participants dimension' (Figure 3.3) both of Case B's technologies imply higher barriers for new entrants, at least in their niche of producer-dyed fibres (as can be exemplified by barriers 6 and 9 in the model).

The analysis of industrial competitors shows some evident advantages of Case B's dyeing process, mainly in what Wilkinson (1987) called the 'breadth of activities', as it enabled the firm to aim towards its present generic business strategy of differentiation-focus at the same time that its association with the resin treatment (and its intrinsic purchasing policy) led to a substantial cost advantage over industrial competitors.

These comments about the relationship between technology and competitive forces are illustrated in Figure 6.10.

In spite of all Case B's advantages, the acrylic-fibres industry is a very mature sector. The management of technology in this business reflects this maturity as their R & D is apparently totally dedicated to end-use development and incremental innovations (this assertion can be exemplified by its indifference to micro-fibres). In terms of corporate strategy this maturity is also recognized but in spite of its profitability, there are repeated rumours in the market-place that Fibrecomp is interested in selling the business (the same rumours suggest that only a failure in price agreement avoided its sale to an Italian group a couple of years ago).

Notes

1. Also during the course of this research a sixth business within the Fibres and Films division (Carbon Fibres) was divested.
2. The sale of the wood pulp business was very representative of Fibrecomp moving towards specialties. This business was profitable (return on capital employed was respectively 40 per cent, 28 per cent ,19 per cent ,42 per cent and 42 per cent in the years from 1984), however, it has a completely distinctive technology basis. Wood pulp is the main raw material in viscose (another of Fibrecomp's businesses); nevertheless the reasons for divesting seem to be linked to the absence of any real synergy (see Section 1.6).
3. The textile group, as well as the wood-pulp business, was consistently profitable; however, it had a completely different technological basis as well as distinctly different marketing forces.
4. The analysis of its efficiency is, however, outside the scope of this research.
5. In the discussion about the movement towards specialization, it can be observed that while in the 1985 *annual report* there were no more than two paragraphs about R & D in its 'review of operations', in the last few years it has taken at least one whole page. At the same time, its location in the report moved from pages 20–4 in 1986–9 to page 18 in 1990 and page 10 in 1991, where it now opens the 'review of operations'. While such observations cannot be related to their real effort towards research intensity, it is certainly a clear indication of the impor-

tance of R & D in relation to the image to be transmitted to the shareholders.

6. It is necessary to avoid the mismatch between the three different levels (macro, meso and micro) of technology strategy analysis adopted in this research methodology and the three organizational levels of research and development (1, 2 and 3) created by Fibrecomp. These two ideas are completely different and there is no relation between the different concepts.

7. Essential to the understanding of some of the present nuances in their markets.

8. Other important fibres of smaller volume production are:
   – Polypropylene, which is used in carpets, upholstery, packaging, sacking, webbing and other industrial processes. It can be used alone or in blends with other fibres for some types of apparel. It is also chemically resistant and strong.
   – Elastane is a yarn based on polyurethanes, and its application derives from its stretching capability. Unlike rubber, it resists perspiration and has an elevate ratio of lightness/performance. It is used mainly in foundation garments and swimwear.
   – Aramids and carbon fibres are advanced fibres used mainly in industrial applications. Their main characteristics are their strength/weight ratio, which can be five times larger than steel. Their role in the textile industry can be disregarded.

9. The present MFA-IV is supervised by a GATT committee, the 'Textile Surveillance Body'. It has 54 participant countries and applies only to trade between developed and developing countries. The basic objective of the MFA was stated in 1974 as: 'To achieve the progressive liberalisation of world trade in textile products, while at the same time to ensure the orderly development of this trade, and avoidance of disruption in individual markets and products' (Silberston, 1991, p. 62).

   While the first part of its objective seems quite far from completion, the second seems to have had enormous importance in protecting the clothing industry of the developed countries from the threat of 'low-cost countries' exportation. Just to give an idea of its importance, Silberston (op. cit. 68) says 'it was estimated that average clothing prices (from all sources) in the UK market might fall by 5 per cent at the retail level, if the MFA were to go'.

10. Unlike nylon fibre producers, who can buy polymer in the form of chips from any nylon producer and perform only the fibre manufacturing.

11. There are six different solvents being used in the commercial production of acrylic fibres: dimethylformamide, dimethylacetamide, sodium thiocyanate (aqueous), zinc chloride (aqueous), dimethyl sulfoxide and ethylene carbonate.

12. To understand the crimping process, it is necessary to go a bit further back to the spinning process. Since the 1960s bicomponent fibres have become a usual product. Such fibres have two layers of polymers, both polyacrylonitrile, although with different amounts and types of additives. The object of spinning a fibre with two 'different components' is to obtain a differential degree of shrinkage or swell between the two layers. As a consequence, when the fibre is submitted to certain conditions of temperature and moisture, it will get a spiral or helical form. As an end-result, the yarns made from such fibres will possess much better handling, simulating the comfort of natural fibres such as cotton and wool.

13. A small amount of continuous filament is also produced to be used in carbon-fibre manufacturing. Some producers also commercialize 'tops' in which the tow has not been cut but broken.

14. Also during this period the Spanish unit was 'demerged' and is now an inde-

pendent Strategic Business Unit (SBU), directed mainly towards the Mediterranean market.

15. While this statement had the agreement of the managers interviewed, they declined to give an estimate of its numbers or the percentage of Case B's turnover applied to its R & D.

16. In the fibre market–place two main units are used in relation to a fibre's average diameter: the old 'denier' and its 'successor' the 'decitex'. The former is the number of grams that 9,000 metres weighs; the latter is the number of grams for each 10,000 metres.

17. Micro-fibre fabrics, owing to small spaces in their weave, allow perspiration to escape but not water drops to get in.

7 Case-study C (adiponitrile/HMDA/ nylon producer)

7.1 Introduction

Case C is a strategic business unit within the Materials division of a company which will be called 'Chemicomp' for the purposes of this book. Chemicomp was incorporated in the mid-1920s, and presently it is a very large, full-range international chemical company with more then 130,000 employees world-wide and an average turnover over the last three years of UK£12 billion. It has seven business groupings which operate world-wide: Agrochemicals and Seeds, Explosives, Industrial Chemicals, Paints, Pharmaceuticals, Specialties and Materials. In addition, there are some regional businesses (not included in those seven) with responsibility for regional management in their locality. Chemicomp has a chairman in each of its three main markets: Europe, North America and Asia-Pacific. The regional chairman supports the chief executive officer (CEO) of each business segment in achieving their business objectives in the region. The resulting structure is matrixial. In several countries the businesses are conducted by a subsidiary company in that particular country. However, international businesses operating in each territory are directed by their own CEO. Major corporate functions are centralized and offer such services to the different strategic business units and subsidiaries as engineering, management services, a group environmental laboratory and a group toxicology laboratory.

Case C has some 8,000 employees and a turnover of around UK£700 million. During the course of this research Case C was moved from the jurisdiction of one subsidiary (responsible for petrochemicals in Europe) to a new 'materials' division. Figure 7.1 represents a simplified view of the whole company.

Case C is an international business responsible not only for production, commercialization, research and development of nylon 6/6 (or polyamide 6/6) but also for several of its chemical precursors such as cyclohexanol/

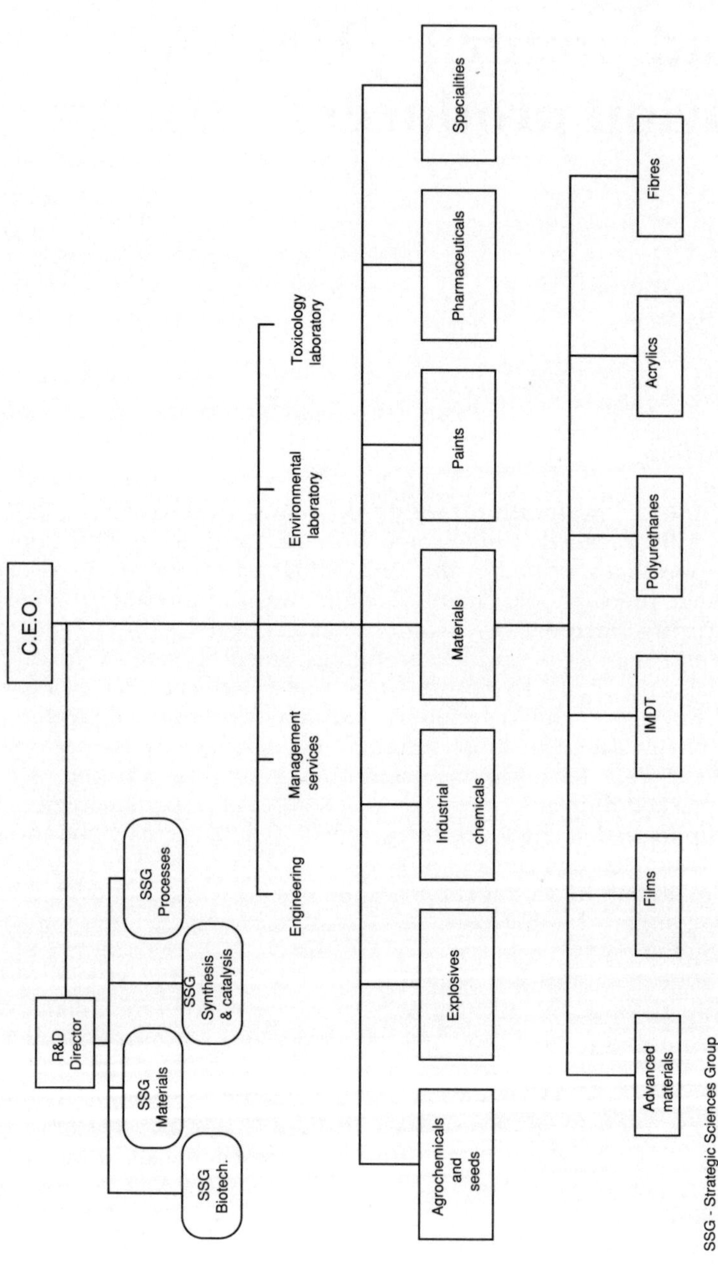

Figure 7.1 Simplified view of Chemicomp

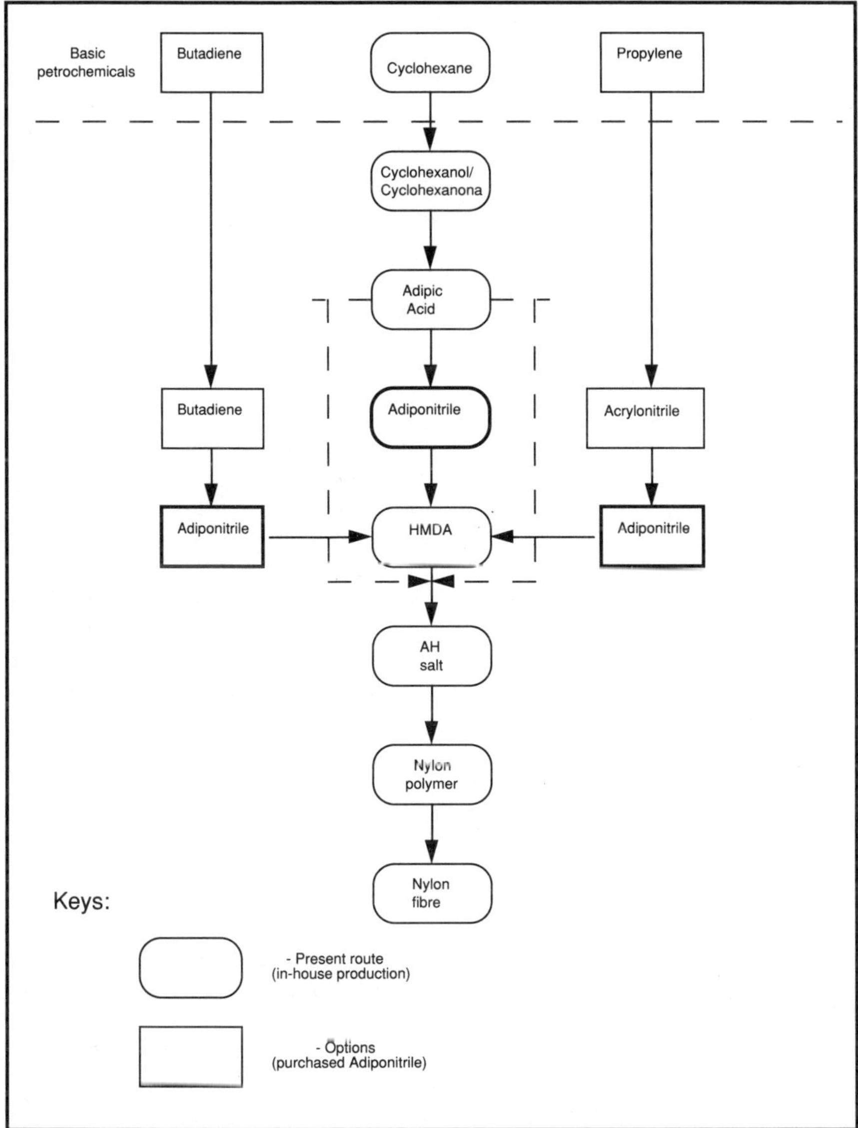

Figure 7.2 Main industrial routes for adiponitrile production.

cyclohexanone, adipic acid, adiponitrile, hexamethylenediamine and AH salt. Their feedstock (cyclohexanol/cyclohexanone) is produced from cyclohexane purchased from one of the other businesses in Chemicomp's industrial chemicals division.

Some 80 per cent of Case C's polyamide is sold as fibre, while the other 20 per cent is sold as polymer (10 per cent) and resin (10 per cent). Case C also sells adipic acid. This combination creates a wide range of products with distinct markets (reviewed in Section 5.4.4), but in spite of this range of products, their main product remains fibre. In this market, Case C is the largest European producer with a capacity estimated at 200,000 tonnes per year.

Case C is not a direct consumer of acrylonitrile but seems to be progressively switching its consumption of hexamethylenediamine (HMDA) from adiponitrile (ADN) produced captively from adipic acid to purchased ADN produced from acrylonitrile and butadiene (see Figure 7.2). Independent of the nature of its future feedstock, however, this product has an extremely important role to play in Case C's technology and business strategy (as will be shown below) and for this reason it was included in the present analysis.

7.2 'Chemicomp's' strategy analysis

As seems to be the case with most of the large chemical companies, Chemicomp's present strategies have their origins in the mid-1970s, the time when the world chemical industry suffered its biggest recession. Since then, Chemicomp has become slimmer, more international and less dependent upon bulk-chemical commodities. As with many other companies, Chemicomp became aware of cyclical down- and up-turns in commodities markets. Chemicomp changed rapidly, and during the 1980s was seen almost as a model of large-company revitalization. During this period Chemicomp actually considerably reduced its dependence upon commodity chemicals. According to Cookson (1991) their executives even thought about getting out of bulk chemicals altogether. The chemical boom of 1985–8, however, brought a strong performance for such commodity products; at the same time some of the new 'higher-value products' ran into commercial difficulties. The main investments and divestments made by Chemicomp from 1985 are listed below (revealed values are also listed, Chemicomp *Annual Reports* 1984–90).

- 1985, purchase of important specialty chemicals/performance plastics, US$750 million.
- 1986, purchase of a large paint manufacturing company in the United States, US$500 million. (This purchase made Chemicomp the largest paint manufacturer in the world.)
- 1987, disposal of its oil and gas subsidiary (continued until 1990) – turnover of UK£1.2 million in 1985.
- 1987, disposal of its participation (37 per cent) in a basic petrochemicals complex.
- 1987, purchase of a large chemical company (US $1,695 million) followed by the sale of all of that company's business in non-specialty areas (compounding a net value of US$800 million to the investment).

From 1987 Chemicomp kept its pace towards higher added-value products through a more punctual strategy, as follows:

- 1988, 30 investments in new businesses and 15 divestments.
- 1989, 22 investments and 26 divestments.
- 1990, 16 investments and 21 divestments.

The overall consequence of all these investments and divestments can be briefly illustrated by the changes in shares of the different divisions upon the Chemicomp global turnover from 1985 to 1990 as follows:

- Oil and gas: from 12 per cent to 0 per cent.
- Pharmaceuticals: from 7 per cent to 11 per cent.
- Paints: from 6 per cent to 13 per cent.
- Other effect products: from 9 per cent to 18 per cent.
- Petrochemicals and plastics: from 18 per cent to 25 per cent.

In the same period, the largest changes in profits share were:

- Pharmaceuticals: from 23 per cent to 47 per cent.
- Paints: from 4 per cent to 10 per cent.
- Agriculture: from 20 per cent to 11 per cent.

Such changes show the efficacy of the implementation of its strategy. Its efficiency, however, has been questioned owing to its performance in terms of profitability on 'effect products'. While this segment had its share of corporate turnover increased from 9 per cent to 18 per cent, its participation in profits changed (over the same period of time) from 1 per cent to 0 per cent.

The present recession, together with the non-existence of changes to the profitability of the specialty-chemicals division, raises questions about the correctness of that strategy. It is not yet known if the company's overall good performance in the last few years was because of the boom in the chemical industry or if their strategy was really paying off. Independent of the answer, Chemicomp has recently launched a new restructuring programme in which the group will focus more selectively on its global strengths. According to Lorenz (1991), investment capital will be provided only to those businesses which have the potential to be strong and competitive on a global basis. 'Other businesses will be managed to extract their cash, or sold' (this idea is of central importance to the present analysis and will be discussed again at the end of this chapter). Also within this restructuring, Chemicomp is redirecting its management towards a higher concentration on strategic decisions and even financial control.[1] One of the clear consequences of this change can be seen in the reduction in numbers of business areas by amalgamation of them. From 14 business areas they now have only seven. This was followed by an equivalent reduction in executive directors on the main board.

7.3 'Chemicomp's' R & D structure and strategy

Chemicomp invests heavily in R & D (around UK £700 million per year), representing more than 5 per cent of its turnover, an investment which has grown consistently and quickly. In absolute terms this growth began during the early 1980s from UK£200 million. Relative to its turnover, such growth began to be consistent in 1984 when, according to Leadbeater (1991), it represented 3 per cent (presently around 5.5 per cent). Almost at the same time that these figures started their strong growth, the company significantly changed its research strategy by largely dropping its programme of corporate-conducted research.[2] In contrast with all its main competitors (at least those of its size), all its research is presently conducted within its businesses. In terms of strategic research not related to present businesses, Chemicomp's only effort is conducted through four 'science strategy groups' which are related to catalysis, biotechnology, materials and processes. These groups (represented in Figure 7.1) are formed solely by senior research managers meeting four times each year (Cookson, 1991). While this type of committee can keep Chemicomp with 'one eye' on strategic subjects, it cannot do much more. It seems that Chemicomp is, in principle, happy to be just an observer of strategic areas not directly linked with its businesses.

In physical terms, there are some exceptions to the straight linkage between businesses and their individual research functions. Chemicomp has at least one common research centre, in the north of England, which is mostly concerned with materials. It employs some 1,000 researchers and was created at a time when materials did not yet exist as a business area. At that time the idea was to explore the potential synergy existing among the different businesses in some way related to materials. Presently, in spite of the existence of a so-called 'materials' area, this centre can still be seen as a conglomerate of different business research groups, an assertion which can be justified by the fact that individual groups (inside that common building) are formed from employees of different businesses, each with its own budget. (This 'business orientation' goes even further, as is discussed below.)

7.4 Market review

This market review will cover only those products relevant to this research; as a consequence products such as adipic acid and cyclohexanol/cyclohexanone will not be analysed.

7.4.1 Adiponitrile/hexamethylenediamine (HMDA)

Adiponitrile and hexamethylenediamine are not commercialized by Case C, and for this reason this product's market review will not be fully extended. There is, however, a very important aspect of the market which is relevant in analysing this sector. This aspect concerns the relative prices of butadiene and acrylonitrile. As shown in Figure 7.2, these two products can be used as raw materials for 'adipo/HMDA' manufacturing. Figure 7.3 shows the fluctuations in these products' market-prices from 1987 to 1991.

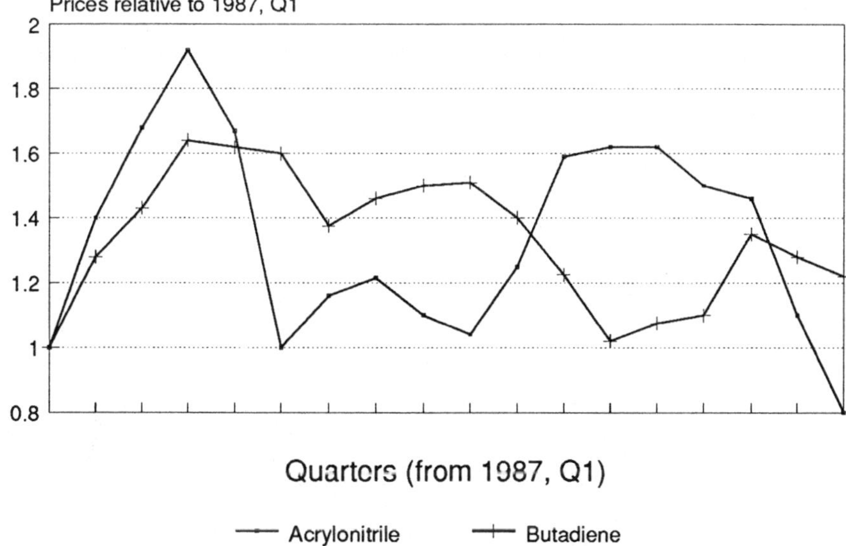

Figure 7.3   The fluctuation in relative prices of butadiene and acrylonitrile from 1987 to 1991.

Considering that raw materials represent between 40 per cent and 75 per cent of the final product cost for a fully depreciated plant (see PEP Year-book, 1989), it is easy to perceive that such an effect can be as important on competitiveness as the process technology itself (this observation is important for an understanding of the delay in Case C strategic decisions discussed below).

7.4.2 Polyamide (or nylon) fibres

Polyamide fibres are usually defined as manufactured fibres in which the fibre-forming substance is a long-chain synthetic polyamide of which less than 85 per cent of the amide (-CONH-) linkages are attached directly to two aromatic rings. Because the definition is expressed as a function of the linkage between monomers instead of the nature of these monomers, the definition is applicable to a wide-ranging group of polymers such as nylons 6, 6/6, 6/12, 6/11, 6/9 and aramids (aromatic polyamides). Nylons 6 and 6/6 together represent approximately 97 per cent of total nylon fibre production.[3] They are, however, quite similar in most of their properties and applications, and it is usual to see their statistics as a total. Most of the comments and statistics below refer to the total of both (unless stated otherwise).

Several aspects of the nylon fibres market are similar to those of acrylics (such as the increasing importance of the Third World on consumption and the different 'periods' in the European market). These can be seen in Section 6.4.1, where a general overview of fibres is presented.

Table 7.1 Nylon fibre consumption by major markets in the United States, Western Europe and Japan (%).

Year	Market	Europe	Japan	USA
1981	Apparel	33.9	27.5	17.7
	Home furnishing	48.8	11.7	63.8
	Industrial	12.6	60.7	18.5
1983	Apparel	30.0	26.5	NA
	Home furnishing	51.1	14.7	NA
	Industrial	14.2	58.7	NA
1986	Apparel	26.5	26.5	12.6
	Home furnishing	52.2	15.4	73.9
	Industrial	16.0	55.1	14.4

Table 7.1 represents nylon-fibre consumption by major markets in Japan, Western Europe and the United States. As can be observed, consumption is quite varied in the different regions. Such a characteristic is the opposite of that shown by acrylic fibres and is due to the greater versatility of polyamides.

The main characteristics of nylon fibres are: its abrasion resistance, which makes it ideal for carpets; its low specific gravity and its stretch recovery, which makes it ideal for hosiery.

Some of the main nylon end-use consumptions are in sheer hosiery/intimate garments and carpets. In this second market (floor covering), nylon's main competitor materials are polyolefin fibres. According to Davies (1987), polyamide participation in this segment fell from 50.5 per cent in 1977 to 42.2 per cent in 1984, while polyolefins grew from 10.5 per cent to 26.2 per cent in the same period in Europe.

Figure 7.4 shows the fibre consumption in the European Community for the production of textile-floorcoverings, illustrating the threat from

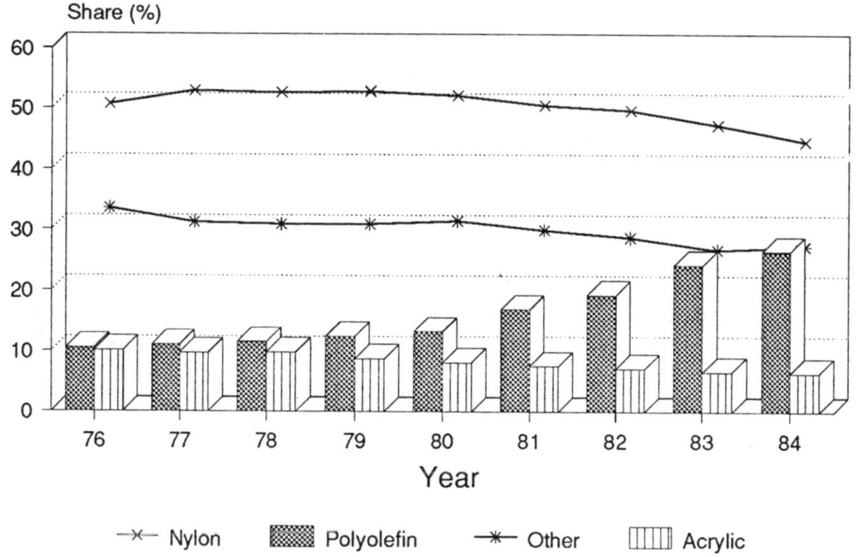

Figure 7.4 The competition with polyolefins (adapted from Davies, 1987).

polyolefins. As can be observed, shares of all other materials are also falling. It is important to highlight, however, that most of this substitution process occurs in carpet and rug backings and is mainly due to polyolefin's lower cost. In spite of some loss of its share, nylon is expected to maintain its presence as a face yarn. Carpets and rugs represent some two-thirds of polyamide fibres in home furnishing, while fabrics are responsible for the other third. In the industrial market, tyres are the major end-use, although consumption is again reasonably well spread through different end-uses.

In comparing the markets for acrylic and nylon fibres, it is clear that nylon's versatility makes its market much less concentrated than acrylics. This distribution is not just in terms of major markets (apparel, home furnishing and industrial) but also within each of these major markets

An important observation is that the main source of growth in nylon markets is in resins, not fibres. This means that beyond the distribution of the nylon-fibres market, there is also a healthy distribution of the basic polymer consumption between fibres and resins. Fibres represent 85 per cent, 73 per cent and 62 per cent of nylon utilization in the United States, Western Europe and Japan respectively, being the complement to resin production.

Case C's overall production breakdown is 10 per cent polymer chips, 10 per cent resins and 80 per cent fibres. Its fibre production is approximately 50 per cent apparel filaments, 30 per cent carpets and 20 per cent industrial filaments. Case C's main threat seems then to be its relatively high cost of production (compared to polyester or Du Pont nylon fibres).[4]

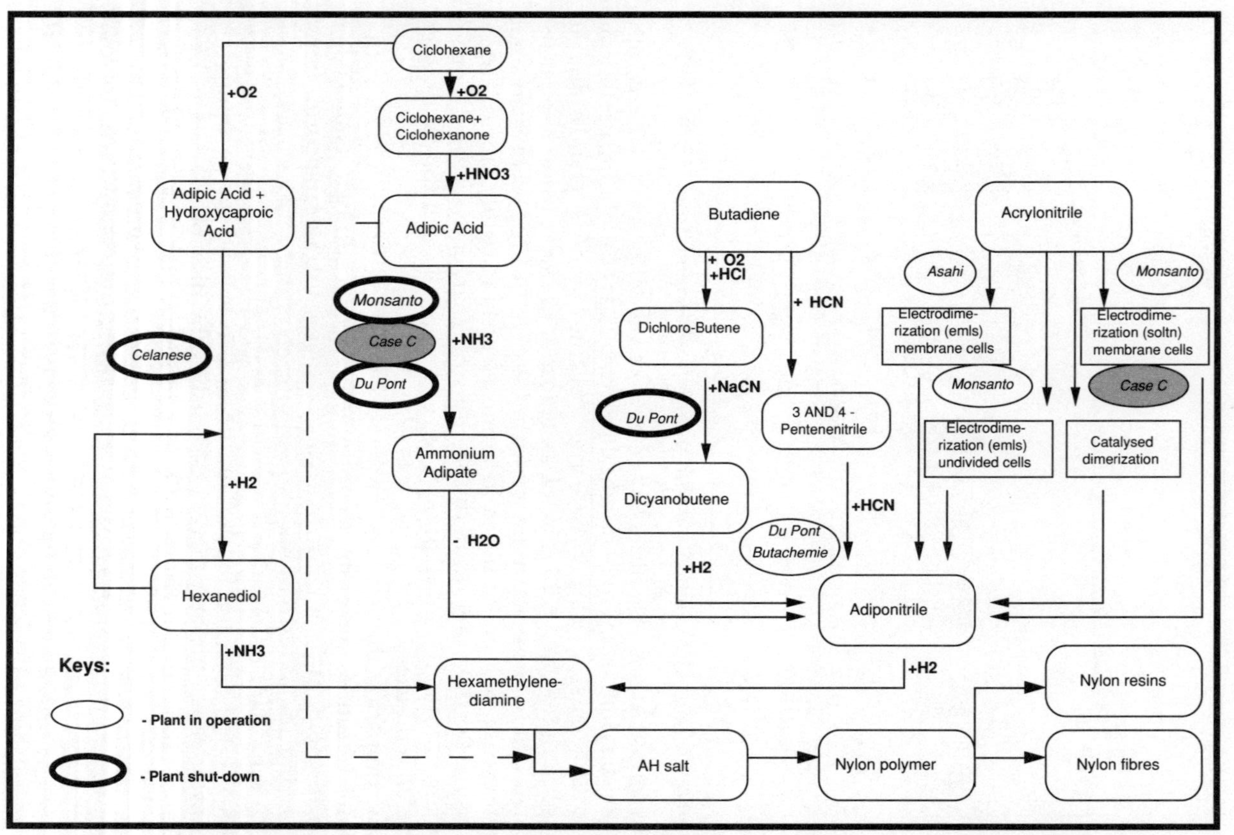

Figure 7.5 The different industrial processes for adiponitrile/hexamethylenediamine production.

7.5 Technology review

As described above, Case C produces from cyclohexane successively: cyclohexanol/cyclohexanone (or KA oil), adipic acid (AA), adiponitrile, hexamethylenediamine, nylon salt, nylon polymer and nylon fibres. The processes to manufacture adipic acid and nylon salt are not critical in terms of technology or even business strategy. Adiponitrile and nylon polymer, however, are specifically relevant to this research. Both provide clear examples of the roles of technology and business strategies as intimately related areas of decision-making and interaction in the chemical industry.[5]

7.5.1 Adiponitrile

All adiponitrile industrially produced throughout the world is used as raw material for hexamethylenediamine production. For this reason they are put together in this short review.

There are four main routes and eight known processes for industrial production of adiponitrile/hexamethylenediamine (illustrated in Figure 7.5). The routes are:

(1) Catalytic oxidation of cyclohexane at high pressure to obtain a mixture of adipic and hydroxycaproic acid, followed by its esterification with recycled 1–6 hexanediol and hydrogenation. The purified diol then reacts under very high pressures with ammonia to produce hexamethylenediamine. This route was used by Celanese in the United States, the only company to use such a process; Celanese subsequently withdrew from this business early in the 1980s (buying nylon salt for its nylon-resin business).

(2) The second route was the main process for several years and was used by different companies, including Case C. In this process (still in use by Case C), a mixture of cyclohexanol and cyclohexanone (obtained by catalytic oxidation of cyclohexane) is oxidized with nitric acid in the presence of a vanadium catalyst to produce adipic acid, which is converted with ammonia (over a dehydrating catalyst) to adiponitrile.

(3) The third route was used in two different processes. The first one (through dicyanobutene) evolved into the second with considerable energy savings. This latter process involves direct cyanation of butadiene, passing through an intermediate mixture of pentenenitriles and butenenitriles. This evolved process also represented an advance in terms of environmental problems, although there is still concern in this respect in relation to the new 'three-steps process'. Du Pont first used the new process in 1971 and converted its second plant in 1983.

(4) The fourth route involves dimerization of acrylonitrile and has several variant processes such as electrodimerization in solution, in suspension, in divided or in undivided cells.

A fifth, and radically different process of acrylonitrile dimerization would involve a catalytic instead of an electrolytic reaction. Its key patents belong to Case C; nevertheless it has not yet been industrially explored (this subject will be further discussed in the section on technology-strategy analysis).

7.5.2 Nylon and nylon fibres

Nylon 6/6 is produced by a condensation reaction between hexamethylenediamine and adipic acid (at this first stage it is called nylon salt). This reaction is conducted in a solution of water or alcohol. All the main producers conduct this step through a continuous process, without considerable differences in terms of technology or yield. Polymerization is the next step and can be made through considerably different technologies. Such differences are not just in processing (continuous or discontinuous) but also in an almost infinite combinations of additives. These additives can determine some of the characteristics which will differentiate one fibre from the other (or one engineering plastic from another) several steps later.

The polymerization step involves not only the incorporation of additives but also the elimination of remaining water or alcohol to the standard level of 0.2 per cent in weight. During this step all unreacted monomer should be recycled, producing a monomer-free polymer.

The continuous process[6] for nylon production is relatively recent (1980s). It was introduced by Du Pont and presents intrinsic economic advantages in terms of cost which are helping to keep this company as the world leader in this market. Case C, on the other hand, still uses the discontinuous process.

7.6 Case C's business strategy analysis

The history of nylon in the United Kingdom is intimately connected with Case C. One year after Du Pont announced its invention (1938), Chemicomp acquired a manufacturing licence for its production in the United Kingdom and the British Commonwealth. Its commercial production began almost at the end of the war (1944) through a joint venture between Chemicomp and Fibrecomp (see Chapter 6). New plants were commissioned in 1948, 1955 and 1965. In 1965 Chemicomp bought out Fibrecomp's interest in this joint venture. Case C became a wholly-owned subsidiary of Chemicomp. Fifteen years later, as a response to the difficult situation in which the chemical industry found itself, Chemicomp created a new and huge subsidiary called 'Chemicomp Petrochemicals and Polymers' from the 'amalgamation' of several independent businesses, including Case C. Such 'amalgamation' was an attempt to increase profitability by increasing efficiency and reducing overhead costs. At the beginning of 1991, apparently in order to have a tighter accountability, Case C was again moved into a new business segment called 'Materials', which, besides nylon, involves advanced materials, films, polyurethanes and acrylics (MMA, not polyacrylonitrile). The consequences of such a concentration affects Case C itself, as is shown below.

Case C's present strategic situation is quite complex both from a commercial and technical point of view. Owing to such complexity, Case C is postponing some strategic decisions for several years. In order to be able to understand some of these issues, it is necessary to make a short review of some of its technical features.

As stated previously, Case C buys cyclohexane from other areas of Chemicomp's business. From this basic raw material, it produces successively cyclohexanol/cyclohexanone (or KA, ketone/alcohol), adipic acid (AA), adiponitrile (ADN), hexamethylenediamine (HMDA), hexamethylenediammonium adipate (AH salt) and finally nylon (resins and fibres). In spite of the fact that Case C sells AA,[7] nylon is by far their main product. This route for production of HMDA from adipic acid is, however, quite obsolete as its costs are considerably higher than those of the modern routes.

Presently three different routes are commercially used by its competitors. Two of them are based on electrohydrodimerization of acrylonitrile (used by Monsanto, BASF and Asahi). The third involves hydrocyanation of butadiene (used by Du Pont and Rhone-Poulenc through their joint venture Butachemie). As stated above, any of these routes can be significantly cheaper than the one used by Case C (via AA).[8]

The fact that Case C has two already depreciated plants for AA and ADN means that the cost issue is not so critical (but still higher than competitors). However, such conditions become more important each year, as the plants require more maintenance and the company's and government's environmental policies became more rigorous.

The complexity of this situation is not restricted to the problems above, but applies equally to the characteristics of the main product down-stream from such intermediates. The fibre industry is in itself quite complex since it is positioned in between the chemical and the textile industries (see Section 6.4). In the specific case of nylon fibres the different markets are extremely diverse, ranging from carpets (2,500 denier) and heavy-duty tyres (1,800 denier) to hosiery (7 denier). Notwithstanding this myriad of processes and markets, the Case C process of nylon production adds more variables. As cited above, Du Pont is the world's largest nylon producer (its capacity is three times as large as Case C's). Du Pont uses a continuous polymerization process to transform HA salt into polymer. Case C does the same polymerization using a discontinuous process through 20 or so different reactors (or autoclaves). Such circumstances impose, again, cost disadvantages for Case C, but at the same time they create more opportunities for a business strategy towards differentiation. Such opportunities come from the versatility inherent to the batch process.[9]

In terms of nylon fibres, this versatility has been fully explored. Case C presently has several trade marks for its fibres. At the top end of the nylon market (their main target) they created three trade names: one for apparel and two for carpets. Such products currently represent some 50 per cent of its production (author's estimate). Such a strategy characterizes a 'differentiation' classification according to Porter's (1980–85) taxonomy. This assertion is confirmed by executives interviewed during this research, as well as in published material (see *Textile Outlook International*, 1987).

It can be concluded that Case C has a portfolio of products with completely distinct characteristics and, what is still more important (at least for the purposes of this book), completely different strategies (such as branded and unbranded fibres) and core technologies (such as fibres and intermediate petrochemicals).

7.7 Case C's technology strategy analysis

Within Case C, the research department is formally (and functionally) a research, technology and engineering (RTE) department. Case C has some 200 researchers, half of them graduates. All projects are conducted as 'capital programmes' and under a 'task-force regime'. Within this research strategy, each project has its own team. Each team has a professional project team manager, an accountant and an interdisciplinary team of chemists, technicians and engineers.

In analysing Case C's technology strategy it is necessary to consider the corporation's influence upon it. Chemicomp has a considerable tradition in terms of research and industrial innovation. Throughout the world Chemicomp employs some 13,000 people in research and development, investing more than 5 per cent of its turnover on this activity. This huge research effort, however, varies considerably among the different divisions and businesses. In the pharmaceutical division, for example, the figure is estimated at around 12 per cent, while in the industrial chemicals division and in Case C itself this number is around 2 per cent.

Even the corporation's organizational structure has a considerable impact upon its strategies. As described above, Case C's position in the corporate structure has changed twice within the last five years. As a consequence of these changes, the present portfolio of products in Case C includes basic commodity products as well as final specialties (such as their branded fibres). Such diversity of products has clear managerial consequences, which are discussed again in Section 7.8.

As in any other discussion about Case C, it is necessary to split this analysis into at least two parts: nylon chemicals (involving the chemical precursors of that polymer) and nylon products (fibres and resins).[10] As a justification for such a division, it can be said that according to interviews, Case C invests some 1.5 per cent of its turnover in fibres research and only 0.5 per cent in its chemical precursors.

7.7.1 Macro-level

7.7.1.1 Nylon chemicals

In terms of nylon precursors it is necessary to analyse the chain of products from adiponitrile. Within this part of the chain, which goes from adiponitrile to AH salt, the main technological efforts in the last few decades have been related to the first step, the adiponitrile production process. As explained above, three basic raw materials were used for industrial production of ADN. Historically speaking, the first two routes were those from adipic acid and butadiene, both dating back to the 1950s. In the early 1960s Monsanto patented and commercialized the first industrial process using acrylonitrile as raw material. In the late 1960s Du Pont responded by developing a new process, in 1971 commissioning its first plant using this process (which still used butadiene as raw material, but with considerable economy once it was based on a three-step process and no longer on the old four-step one). At that time this new plant led to Case C aborting its effort to develop a similar process.[11] Since then, two slightly different processes were developed and industrially used by Asahi and Monsanto itself (in the second case with a considerable reduction in energy consumption and, consequently, costs). As can be seen, all producers were striving to reach an independent technology which could enable lower production costs.

7.7.1.2 Nylon products

In terms of fibres, nylon's world-wide market situation presents some similarities to those for acrylics (described in Section 6.4). Such similarities are illustrated by a forecasted growth of consumption in industrialized countries from 1.5 per cent to 2.0 per cent over the next few years, compared to an average of 4 per cent in developing countries. In contrast to acrylic fibres, however, nylon 6/6 world production is almost completely in developed countries. Also unlike acrylic fibres, nylon fibres have a broader range of applications and can be produced from purchased polymer. Possibly owing to that broader range, it is very difficult to identify its common trends (as suggested by Mulder and Vergragt, 1990). On the more selective side of the apparel market a technological trend can be identified towards sophisticated products. Within that trend, specialized products should be produced, aiming at the top of the market in developed countries. In this sense, one of the clearest technological innovations is again the microfibres (see Davies, 1991 and Green, 1992).

7.7.2 Meso-level

7.7.2.1 Nylon chemicals

While Du Pont's butadiene process could not be 'turned around', the same was not true for the acrylonitrile-based processes. Just a few years after Case C gave up butadiene technology, an academic publication about catalytic dimerization of acrylonitrile triggered an enormous effort by that company to develop a new process. From the mid-1970s to the beginning of the 1980s, Case C developed a completely new process to produce adiponitrile from acrylonitrile in which AN is dimerized not electrolytically (as described above) but through a catalytic process. From 1977–8 this project became collaborative research with another major ADN producer.

Case C invested heavily in this new process until 1980, at which point the periodical down-turn in the chemical industry hit Chemicomp with particular strength. Case C reduced its R & D efforts to a minimum, and most of the work was conducted by its 'research partner'. In accordance with this agreement, the results and efforts of this research were shared to a certain degree. From 1984 (with the improvement of its financial health) Case C resumed the project, which was completed in 1990. Both companies own patents related to the process. The full content of this collaboration agreement, including its commercial counterparts, was not revealed, but most of the patents belong to Case C.

7.7.2.2 Nylon products

Within the generic trend towards specialization (in developed countries' industries) Case C is trying to differentiate itself through a combination of 'end-use development' and applied research, supported by strong marketing and customer technical support. Based upon this combined effort Case C created the three trade names referred to above. Case C has 12 different trade names in fibres. The last three, however, have different purposes and are related to a new strategy (differentiation as cited above). Instead of using the trade names for a specific product (as in the past), Case C uses the same trade names for several different products. Instead of trying to link their trade-mark with one product, they are trying to link the mark with a group of characteristics. Such characteristics are related mainly to quality, not just technical quality in a statistical sense, but in a way in which it can be assured that the product is absolutely right for its applications. In order to be certain that one product is ideal for the intended purpose, Case C may need, for example, to modify its polymerization process for some batches. Such changes may involve not just the reaction conditions (such as temperatures, times, fluxes) but also chemical modifications in the polymer (for example, adding chemical modifiers such as polymer end-groups which attract specific dyes or other additives). Other easier modifications can be achieved with the additives used after polymerization or related to specific shapes to the fibre. As can be deduced from Sections 6.5 and 7.5, there are infinite combinations to be made with all the different variables involved in fibre manufacturing.

7.7.3 Micro-level

Technology-strategy generation in Case C has its beginning at the corporation level and, at least at this level, is managerially sophisticated. The generation of strategy seems to have its first 'bricks' settled during the annual budget proposal. At that time each SBU presents its proposals, including research investment. Each proposal should then pass through the business area CEO and his advisory board, where it may eventually be slightly adapted to a broader budget and strategy. This process is repeated again at the level of the group's board of directors. At this point it is analysed and discussed not just with a business view but also through a functional perspective. The functional review of research and development is then compared with competitors and with the overall business and financial requirements. In the next step, the up-down process begins, and the ideas flow again from a conceptual and generic view to practical and detailed objectives. Part of the sophistication referred to above is exemplified by the use of expert decision-aids internally adapted or developed for strategic analysis of research and development (see Coombs, 1991, p.168). Such decision-aids are used mainly at corporate level where Chemicomp's enormous size creates difficulties for an equilibrated analysis of all its different technology demands.

At the business level, this case-study again shows some sophistication in its management of technology. The first point for discussion is the existence of a 'business planning' manager (as in Case A above) formally in charge of the linkage between research/technology and business strategy. Unlike Case A, however, this executive has a research background. The second point is that Case C, as with all other SBUs in Chemicomp, has a periodic technology audit (see Fusfeld, 1978) conducted by a senior corporate manager with an exclusive audit function. This exercise involves the participation of senior researchers, technical assistance, business planning, marketing and production managers. The exercise take some two or three days of exclusive dedication away from the working site. During this time subjects such as present and future technological capabilities are extensively discussed.

7.8 Conclusion

Chemicomp's strategy towards speciality products and away from commodity chemicals proved to be relatively effective and efficient until the late 1980s. However, since then this has been put in doubt; in spite of these doubts, Case C was clearly affected by that strategy. Their 'business strategy' involving branded fibres is easily identified as 'differentiation' and is evidently an effort to direct its products to a 'less cyclical', 'premium-payer' end-market as truly speciality products.

The R & D organizational structure shows the virtual absence of corporate long-term research and a strong dependence by research activities upon the individual businesses management. Such a structure steers the research trends towards short-term applied efforts and end-use development which seems to suggest an absence of synchronism with a corporate strategy towards high-value (high-technology) products. Such an observation, however, cannot be conclusive as internal communication channels between the individual R & D groups possibly provide partial compensation for the absence of corporate central laboratories.

Another important corporate movement is clearly directed towards concentration of managerial powers. This last assertion can be exemplified by several mergers between businesses and business segments. The effects of such concentration upon Case C are, again, quite clear and can be illustrated by the unified management and accountability of 'nylon chemicals' and 'nylon products'. The analysis of this SBU proved to be quite difficult. In order to conduct such an analysis, it was necessary to split each section in two parts (nylon chemicals and nylon products). This procedure shows the difficulties involved in the management of such an SBU. This 'unnatural' amalgamation of businesses (which are so diverse) is a consequence of the corporate managerial tactic towards concentration and seems to make sense when put together with Lorenz's (1991) comments about their strategies cited in Section 7.2: 'and investment capital will be provided only to those [businesses] which have the potential to be strong and competitive on a global basis.'

As described in Section 7.3, Case C's costs are considerably higher than some of its main competitors. Such a situation is quite uncomfortable and, combined with environmental problems, makes a radical change necessary. As described before, Case C's main alternatives are:

(1) to buy adiponitrile produced from acrylonitrile,
(2) to buy adiponitrile produced from butadiene,
(3) to build an adiponitrile plant based upon their proprietary technology (acrylonitrile catalytic dimerization).

Option 1 has some cost disadvantages as the route from butadiene seems considerably cheaper for the next few years (such a conclusion is not 'clear cut' because of the fluctuations in prices illustrated in Figure 7.3. Nevertheless, it seems quite probable, for reasons beyond the scope of this research, that such prices should keep the butadiene route attractive).

Option 2 implies buying from one of its main competitors. This option is strategically embarrassing as they would, in effect, be sustaining their competitor's profitability, which would support their R & D and help to keep the competitor's leadership in production costs.

Option 3 could be the best solution, nevertheless it would involve large investment capital.

Choosing option 1 above, Case C could be in line with environmental policies but would not be optimizing its profitability. By choosing option 2, it is possible to foresee the optimization of profitability but also future environmental problems allied to a very weak strategic position. Option 3 would be the best option if not requiring massive investment.

The observation about investment is the connecting point with Lorenz's (1991) conclusions. It seems also to be associated with the merger of 'nylon chemicals' and 'nylon products' as explained below.

The fibre business has been marginally profitable.[12] The best way to improve its profitability and to guarantee that it is in line with Chemicomp's environmental policies involves the need to invest heavily in a novel technology. This option, however, implies certainty about its competitiveness world-wide which is not dependent solely upon adiponitrile technology but on several other technological and economical issues described in this chapter. As the production of nylon chemicals within Case C

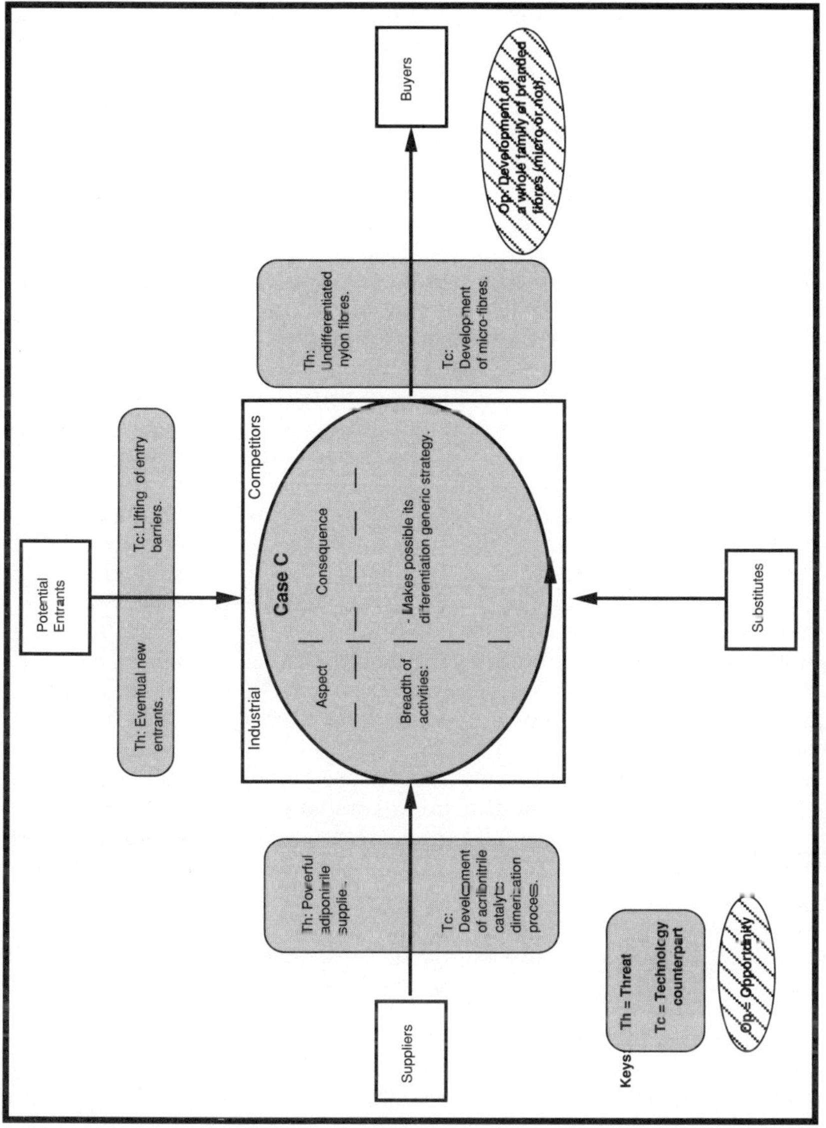

Figure 7.6 Technology and the five competitive forces (Case C)

(at least on the present scale) is not viable without nylon products (and vice versa), it is probable that the merger of the different businesses was already one step in the decision-making process about its continuity (as a whole) within Chemicomp's business portfolio.

At the macro-level a 'natural trajectory' towards cheaper raw materials for nylon chemicals was observed, evolving from adipic acid and the four-step butadiene process towards butadiene (three steps) and acrylonitrile (several processes). A common aspect in the main nylon producers' strategies is their struggle to develop independent proprietary technologies. In terms of nylon products themselves, it was not possible to see any natural trajectory, as even batch processes still compete with continuous ones.

At the meso-level of analysis it is obvious that Case C's initial plan was to develop an independent competitive technology, first from butadiene and later from acrylonitrile dimerization. As observed by Mulder and Vergragt (1990), this was a consequence of the opportunity (provided by an academic paper) combined with the company's internal capability and market situation. Such a combination of factors would lead Case C to differentiate itself by using a catalytic process instead of an electrolytic one.

In terms of fibres, Case C's technology strategy is directed towards the exploration of the intrinsic flexibility of its process. Such a strategy is, however, a mixture of technology and business strategy: it being virtually impossible to separate one from the other.

At the micro-level it was possible to identify the use of some techniques and decision-aids aimed at helping with the corporate strategy-generation process such as technology versus business quality matrices and technological forecasting (see Cookson, 1991). Also at the business level, it was possible to identify the use of decision-aids and even of a technology audit technique. In all cases, however, the techniques and tools play a role mainly of common-language generation (as described by Coombs, 1991).

It can be said as well that this company has (and it is perceived by its managers as such) a technology strategy and within that an R & D strategy. It can also be claimed that these strategies are generated in a bottom-up/top-down process starting at the middle-management level (where the annual budget starts) and rising to the top levels where it is combined with the overall strategies then returning to the middle level where it takes its material form.

In terms of the models for R & D project generation from business-strategy analysis described in Chapter 3, the development of the catalytic dimerization process can be seen as an R & D counterpart (item A in Figure 3.2) to a potentially powerful supplier (item 1). The development of the branded fibres, on the other hand, can be seen as a R & D counterpart (item B on the right side of Figure 3.2) to the absence of bargaining power from Case C (items 3 and 7). These comments are illustrated in Figure 7.6.

Notes

1. Opposite to what seems to be happening to most other big companies such as BP (see Lorenz, 1990a, 1990b and 1990c).
2. Presently, the only centrally controlled research is related to colloid science and involves 35–40 researchers.
3. In 1990 world production of nylon fibres was approximately 2.3 million tonnes of nylon 6 (83 per cent filament) and 1.5 million tonnes of nylon 6/6 (73 per cent filament). In terms of resins, 420,000 tonnes of nylon 6 were produced against some 380,000 of nylon 6/6. (Author's estimate from analysis of several different sources.)
4. Du Pont is the world leader in this market.
5. For an overview of its technologies see Kirk-Othmer, 1975.
6. While the discontinuous (or batch) processes require successive operations of filling, processing and discharge (often also cleaning), continuous ones are characterized by a 'steady state' in which all conditions are kept constant. Typically the vessels are kept at a constant volume through equivalent rates of feeding and discharge, eliminating 'dead times' such as filling, cleaning, etc. Continuous processes, however, require a higher level of engineering and control than their equivalent batch processes. Such disadvantage is fully compensated in case of high-volume production. Another disadvantage (more difficult to 'turn round') relates to the intrinsic inflexibility of continuous processes. A typical continuous petrochemical process may require several days to achieve the 'steady state' after a condition change (even if perfectly engineered).
7. Case C sells to the merchant market between 30,000 and 50,000 tonnes of adipic acid per year.
8. To a new plant the difference in terms of production cost can be as high as 100 per cent (see *PEP Yearbook*, 1989).
9. See Freeman (1982) for a discussion about the importance, in economic terms, of continuous-process introduction during the first half of this century. Bessant (1991) presents, on the other hand, some discussion about the importance of flexible manufacturing at present and in the future. Both arguments give weight to an understanding of the discussion below.
10. As depicted above, fibres represent some 80 per cent of Case C's sales by volume. In terms of turnover this data was not revealed. It can, however, be expected to be not far from this number.
11. The fact that during the 1960s Case C were already trying to develop a new process shows how obsolete is their adipic acid process at the present.
12. During the last seven years its average share of Chemicomp turnover has been approximately 5.8 per cent, while its average share of profits has been only around 2.7 per cent, presenting some losses in 1991.

8 Case-study D (synthetic latex producer)

8.1 Introduction

Case D is a small company with about 200 employees and a turnover of UK£35 million. The company ownership is divided between two large and completely different companies, each with 50 per cent of its shares. One of the companies is a long-established British financial conglomerate (a public company) with considerable interests in speciality chemicals and a turnover of about UK£260 million. For the purposes of this book this company will be called 'Britco' while the second shareholder will be called 'Cheminc'.

Cheminc is an American incorporated chemical company with a turnover of about US$1 billion and has five key businesses: coatings, adhesives, composites, forest products and paper. It is fully owned by a Japanese large chemical corporation which is a world-wide supplier of synthetic resins, printing inks and pigments and which will be referred to as 'Inkco'.

Within 'Britco' Case D is located in the speciality-chemicals division, while within 'Cheminc' it is in the emulsion polymers division.[1]

Case D was formed in 1963 to broaden the activities of one of its two parent companies (neither of the present ones) at the time producing only natural latex and polyvinylacetate. Since then there have been several changes in the firm's ownership. The last three important changes were in 1975, 1980 and 1987, respectively, as depicted below.

In 1975 Cheminc became one of the partners, and in 1980 Britco (also with large interests in natural latex) bid for another 50 per cent of the company's stock. In 1987 when Cheminc was about to sell its stake to Enichem (one of Case D's main competitors), it was overtaken by Inkco which decided to keep Case D in its portfolio.

Case D's business is defined as the 'design and manufacture of synthetic latex and latex-based compounds for industrial applications the world over'. Its main production site is a latex manufacturing plant in the north-

162

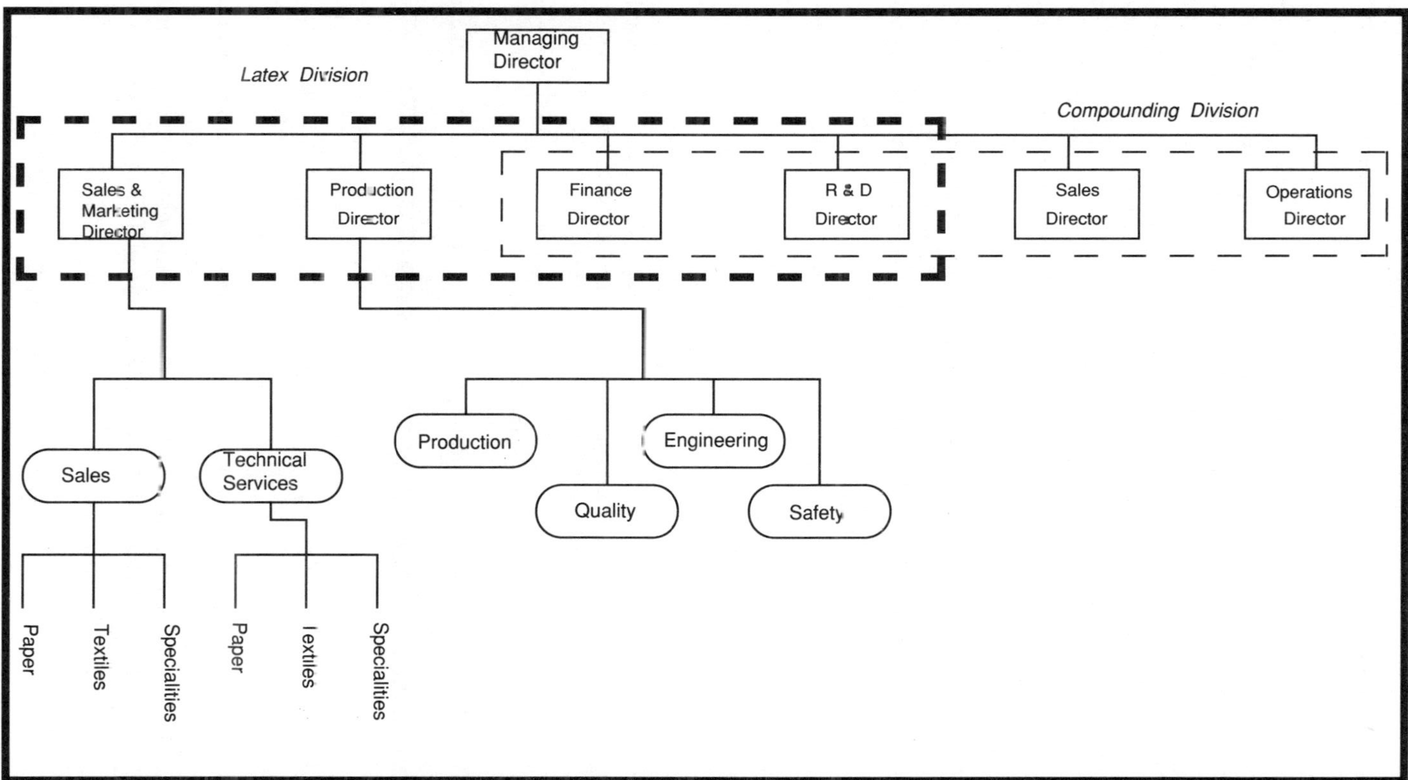

Figure 8.1 Case D's organizational structure.

east of England. Using these latices as their main raw material two other smaller manufacturing plants (one in Scotland the other also in the north of England) produce latex compounds.

In terms of organizational structure Case D has two divisions, one responsible for compounds and the other for latices. The compounding division has the carpet industry as its main market, while the latex division targets three distinct markets: paper and board, textiles and specialities. A simplified view of the firm's organizational structure is illustrated in Figure 8.1, showing a typical departmental organization. The compounds division has some 70 employees while another 130 work in the latex division (this number includes those involved in finance and R & D which are functions shared between the two divisions). The two functions more directly involved with this research are 'sales and marketing' and 'research and development'. These functions have 35 and seven employees respectively.

In spite of its small size, Case D has a range of some 30 different products under regular production and another 30 or so that are manufactured only when ordered by a customer. Its yearly production capacity is around 70,000 tonnes of latex and 40,000 tonnes of compounds.

8.2 Corporate strategies

8.2.1 'Inkco'

Inkco's long-term goal is to be among the largest chemical companies in the world in the 21st century. In order to achieve such a goal it designed a strategy whose main components pass through its internationalization and through a move towards high-technology content/high value-added products. Its main steps towards internationalization were made in 1976 when it bought 'Star' Chemical's graphic arts business and in 1978 when its hostile bid of US$540 million (Peter, 1989) for Cheminc was successful (resulting in full control of that company). At that time Cheminc already operated in 12 different countries (mainly in the important American and German markets). Besides this highly interesting international character, Cheminc had several previous technical-exchange agreements with Inkco before the acquisition and was among the largest American producers in several of the Inkco core businesses such as printing inks, organic pigments and thermosetting resins.

Inkco's thrust towards high value-added products is evident when reading its annual reports as well as the specialized literature on its different individual businesses. The company has been quite innovative in areas where it is a new entrant such as the pharmaceutical industry, food additives, liquid crystal displays and other polymer-related businesses (*Japan Chemical Week*, 1991a).

8.2.2 'Cheminc'

Cheminc's objective for the year 2000 is to achieve a turnover of US$2 billion. In order to do this, one of the components of its strategy seems to be to concentrate successively on core businesses. By the end of the 1980s they defined their five core businesses as coatings, adhesives, composites, paper and forest products. In 1991, however, they announced the divestment of those businesses related to forest products in their Canadian subsidiary (*Chemical Week*, 1991b). This strategic move indicates further concentration around the remaining four areas (no other considerable investment was made in new areas).

It can be argued that in spite of its large interface (in terms of markets as well as in terms of technologies) with Cheminc, Case D's strategic value as part of Cheminc's and mainly Inkco's portfolio is dubious. Support for such a statement comes first from the relatively small size of the company. While the parent company has a turnover of about US$1 billion, Case D doesn't reach US$70 million, representing only 7 per cent of Cheminc sales (and less than 2.2 per cent of Inkco). Second, and more important, is the fact that Case D has a sister company in Germany with almost exactly the same line of products and scope of market. This German firm, exactly the same size and capacity of Case D, is also 50 per cent owned by Cheminc. In spite of the fact that the German company has some disadvantages in terms of environmental concerns (it is located near dwellings, differently to Case D), it operates in a much stronger domestic economy.[2] The importance of the parent company's strategy is quite different when considering the other shareholder as depicted below.

8.2.3 'Britco'

Britco's present assets are linked to two main areas of building and speciality chemicals products. The chemical division, presently responsible for 55.4 per cent of the group's turnover, has all its businesses within five areas: natural latices, synthetic latices (containing three companies), emulsions, consumer chemicals and other speciality chemicals. In spite of the fact that it is clearly in the scope of speciality chemicals, Case D is completely independent of its sister companies (in this side of the group), sharing neither technology nor any other productive or managerial function. Despite this absence of synergy, Case D is an important part of Britco's portfolio on account of its present profitability (as also discussed in Section 8.6). The whole chemical division generated a turnover of UK£147 million and an operational profit of UK£17.8 million during the 1992 fiscal year. This was an excellent result for an industry in deep recession, demonstrating that its strategic move towards high value-added products is paying off.

8.3 Corporate research and development

Inkco invests heavily in R & D. By 1992 this investment reached US$180 million, representing almost 5 per cent of its turnover and involving more than 1,500 employees. As cited above this company has had a considerable number of innovations (both in terms of products and processes). Inkco has several business-orientated research laboratories (seven in 1991) and one corporate research centre. Apparently, there is some overlapping and cooperation among the different centres, as cited in the 1992 report, for example:

- Between the Graphic Arts Centre in Tokyo and the 'Star' Chemical's Carlstad Laboratory in New Jersey;
- Between the Chemical Technical Laboratory in Kashima and 'Star's' Research Operation Center in Cincinnati (organic pigments);
- Between Polymer R & D Centres in Kanto and Kansai.
- Between Cheminc Chemical's Central Research Laboratory in North Carolina and Inkco's Berlin's Research and Development Laboratory Europe.

The parent company seems to try to stimulate regular meetings among researchers at its own international technical conferences (see Sub-Section 8.7.3) as well as personnel exchange among the different centres.

In terms of the other parent company (Britco), as might be expected from a financial conglomerate, the strategies are relatively independent of technological nuances (this aspect is discussed further in the next section). Britco's investment in R & D was quite small. According to its annual reports, during the years of 1990 and 1991 its average expenditure in R & D was only around 0.45 per cent of its turnover, a sharp contrast with Inkco (such facts, however, are not necessarily an indication of the absence of a technology strategy).

8.4 Synthetic latices market review

Before discussing the synthetic-elastomers market it is necessary to clarify its several technical and commercial interfaces and overlaps with some other groups of chemical products such as natural latices, dry synthetic elastomers and polymeric emulsions.

The logical starting point for such a discussion is a definition of polymer: 'a substance having large molecules consisting of repeated units (monomers)'. There are a huge number of different synthetic and natural polymers; among the latter group, natural rubber is the most interesting example in view of the purposes of this case-study, as will be explained below.

Polymers with the special ability to undergo deformation under the influence of a force and regain their original shape once the force has been removed are classified as 'elastomers'. Natural rubber on the other hand is the polymeric substance obtained by coagulating and drying the sap of the tree *Hevea Brazilienses*.[3] The sap is a latex, which is by definition, 'a suspension of very small elastomeric particles dispersed in water'. As happens with the other types of chemicals described above, latices can be also synthetic or natural.

The last correlated class of polymers is the polymeric emulsions which, like the latices, are small polymeric particles dispersed in water. In contrast with latices, however, the emulsions are an homogeneous material where the coalescence and precipitation of particles are avoided, usually by the use of emulsifiers, and the polymeric particles are not necessarily elastomerics.

8.4.1 Elastomers

'Cis-1-4-polyisoprene' was the first elastomer in use by mankind. It had been used for centuries by South American Indians before the English chemist, Joseph Priestley, named it as rubber when he noted that the *Hevea* tree's gum could be used to 'rub out' scribbles made with a pencil. Natural rubber, as it is usually called, can be obtained from more than 200 different species of plants (Kirk-Othmer, 1975), although more than 99 per cent of world production is based upon the latex extracted from the *Hevea Brazilienses*. The invention of the vulcanization process in 1839 (see Chapter 4) made rubber a much more useful material. It improved its properties and provided the necessary conditions for the development of the first practical tyre, a new and promising market. The first statistics about this new commodity were compiled some 10 years later and showed a world production of four tonnes. Presently the world consumption of elastomers is about 15 million tonnes per year consisting of five million tonnes of natural rubber and 10 million of synthetic elastomers. According to the International Institute of Synthetic Rubber Producers (*Chemical Marketing Reporter*, 1991), this production should grow at a yearly average rate of 2.1 per cent reaching a total production of about 17 million tonnes by 1995.

8.4.1.1 Synthetic elastomers

Synthetic elastomers can be divided into two categories: general purpose and speciality elastomers. The first group can be divided further into tyre elastomers including butyl, polybutadiene, polyisoprene and SBR (styrene butadiene rubber) and high-performance elastomers including polychloroprene, ethylene-propylene and nitrile elastomers. Figure 8.2 illustrates this classification within a broader taxonomy of polymers.

As described in Chapter 4, the industrial manufacture of synthetic rubber was pioneered in Germany in 1934. Since then the production of synthetic elastomers has grown at an astonishing rate, achieving approximately 50 per cent of the market and production shares as early as 1960 and peaking to a maximum of about 71 per cent in 1979 when the higher raw-material prices upset the expectation that synthetics would replace natural rubbers altogether. Presently the market share seems to have

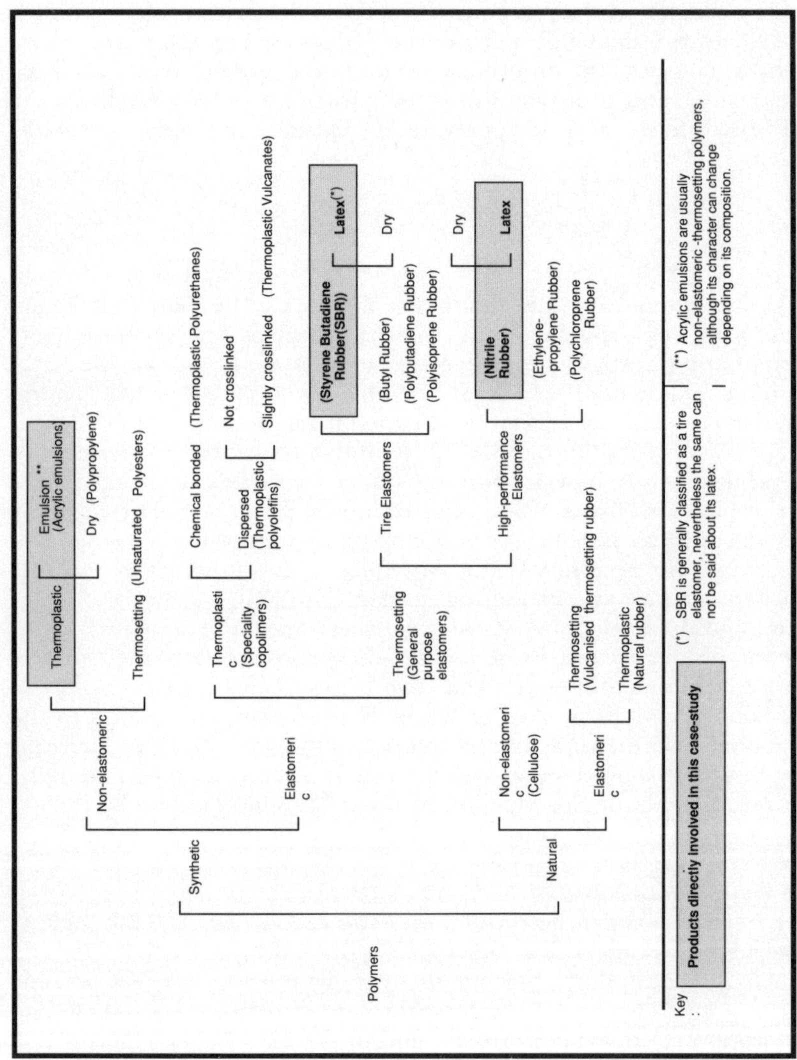

Figure 8.2 Synthetic latices within a polymer taxonomy.

stabilized with synthetic elastomers responsible for two-thirds of world production and consumption of elastomers.[4]

Synthetic elastomer's main end-use is, by far, tyre products, representing some 60 per cent of their total consumption. The second largest application is in the manufacture of moulded, calendered and extruded products (13 per cent of North American consumption and 20 per cent of Western European). All other major uses such as footwear, belting, hose, electrical adhesives and sealants have market shares of between 1.5 per cent and 3 per cent.

8.4.1.1.1 Synthetic latices

Synthetic latices present the same basic molecular composition as their dry counterparts, nevertheless they differ in several other aspects such as processing and compounding but mainly in application. Although latices can be suitable for coagulation (in order to be processed in solid state), their fluid state provides several advantages for specific types of applications such as saturation of porous materials, paper coating or preparation of foam sponges. The fluid state renders advantages in terms of rates of application in continuous processing as well as power savings during compounding processes. Also the environmental hazards, usually associated with flammable and toxic solvents, can be avoided by using latices instead of dry-rubber solutions.

As described in *Rubber and Plastic News* (1991), SBR latices should present one of the largest growths in terms of synthetic elastomers in North America (where carboxylated SBR latex holds a share of 16 per cent of synthetic elastomer production), achieving an average growth of 4.9 per cent on a yearly basis.

8.5 Synthetic latices technology review

As described above, Case D produces some 60 different products, although all of them are based on just three monomers (styrene, butadiene and acrylonitrile). More than 90 per cent of its production is concerned with variations of styrene-butadiene polymers (SBR). The second largest group of latices are those made of nitrile-butadiene polymers (NBR).

The main step in both manufacturing processes is polymerization. During this step the polymer's molecular structure will be defined as well as its core physical and chemical characteristics. In both cases the main component is butadiene. Also in both processes the type of polymerization used is emulsion polymerization, where monomers are emulsified prior to the polymerization. The droplets of monomer are stabilized by the presence of emulsifiers which avoid their coalescence by keeping a uniform distribution of droplets through the continuous medium of water. This type of polymerization was first employed by IG Farben in the 1920s when they produced some industrial SBR but later discontinued. Its first successful use was also initiated by IG in 1934 in production of BUNA-N (a copolymer of butadiene and acrylonitrile). In 1942, when the SBR started to be successfully produced in the United States, the polymerization process was again the emulsion process, which has been employed since then in all industrial production with few exceptions. These exceptions are related to solution polymerization, which arose in the last two decades and is becoming more and more important although still carried out only for special applications. As Case D uses only emulsion polymerization this brief discussion will be focused on this process.

A considerably different type of emulsion polymerization was launched in 1947. The critical difference between these two processes is the temperature. The traditional 'hot rubber' has always been polymerized at approximately 50° centigrade, while the new cold process is at only 5° (using a more active initiator). As a consequence of its lower temperature, the second process yields a product with a lower content of branched and cross-linked polymer chains, resulting in better processability and superior properties. On the other hand it presents intrinsic cost disadvantages resulting from lower conversion of monomers (typically 60 per cent against 70–75 per cent for the hot process). Heightening such a disadvantage is the fact that latices are usually sold at 65 per cent, meaning that a costly concentration must be effected with the polymers produced using this process.[5]

A further evolution from the original synthetic rubbers (mainly on what concerns latices) was the development of the so-called 'carboxylated grades', where the polymerization is conducted in the presence of a carboxylic acid in order to obtain a product with better processability and physical properties more appropriate to coating applications.

Case D produces only random (see Note 5) SB and NB latices through the hot process, both can be carboxylated or not, although some 95 per cent of their products are carboxylated for marketing reasons.

In the hot process the reaction is terminated through the addition of substances which have the property of destroying any remaining initiator in the polymerization reactor (shortstops) and free radicals (chemical species that keep the polymerization going after the initiation). The next step in the process is the removal and recovering of unreacted monomers for recycling. Butadiene is flashed off followed by vacuum distillation and styrene removed through steam stripping. When acrylonitrile is present it is also removed by stripping.

The last step in the manufacture of latex is compounding. Although an easy unit operation, compounding involves hundreds or even thousands of possible combinations of the different additives. Among the most commonly used additives are preservatives (bacteriostats and fungicides); thickeners (for example, methylcellulose, sodium carboxymethylcellulose, sodium polyacrylate, alginates, karaya gum, casein, starch, locust bean gum, silicates, etc. (*Encyclopedia of Plastics and Polymers*, 1976); anti-foaming agents (silicones and others); vulcanizing agents (amines, aldehydes, guanidines, thiazoles, thiurams, dithiocarbamates, sulfur, zinc oxide, melamines, methylureas, etc.); antioxidants (secondary amines, phenol-derivatives, phosphites, etc.); softeners; stiffeners; tackifiers; fillers; deodorants and even stabilizers (surface-active compounds) which are also added during the polymerization. The existence of so many different combinations opens the way to the creation of niche markets and to strategies towards differentiation as discussed below.

8.6 Case D's business strategy analysis

According to the information collected during the interviews, Case D's board is composed of one representative from Inkco, one from Cheminc, one from Britco and two executive directors from Case D itself. As described by the interviewee, requirements made by the parent companies' representatives are mostly of a financial nature. It is understood, however, that in the past they were responsible for the settling of a market geographical division between Case D and its German sister company.

As described above Case D has three distinct markets: paper and board, textiles and specialities. Among the three, the one linked to textiles is the most mature, with very few areas of market growth. In spite of this maturity Case D seems to be committed to the area as can be understood from the existence of a division (kept to proportions) dedicated exclusively to the carpets market. In order to keep this business running, Case D developed a 'very low profile' structure (very low investment and operational costs), although responsive. Still responsible for some 40 per cent of Case D's market in terms of volume (in contrast with a participation of 50 per cent only five years ago), this sector could not be abandoned. It presents several characteristics of a typical 'cash-cow business' in accordance with the Boston Consulting Group's classical taxonomy.

Case D has two sites operating exclusively for the carpet industry. The idea is to support the growing 'just-in-time' requirements of the carpet industry operating locally near the industry's main geographical areas. In order to do so it has a compounding site in the north of England (neighbouring the main carpet industrial sites) with some 60 employees and an independent accountability. The second compounding site (in Scotland) can eventually be operated by two or three operators; nevertheless, it may be enough to assure a fast, close and differentiated service to the local industry.

The paper and board sector seems to be quite symmetrically opposed to the textile one. Since 1987 its participation in company sales grew from 40 per cent to nearly 50 per cent in volume. In the paper and board industry latices are used as a binder agent for coating. Nearly 90 per cent of all printed paper use some type of latex as a bonding agent between the other two main constituents, woodpulp and clay. Printing paper is a flourishing industry, and the consumption of latex for coating is expected to double within the next eight years (*Japan Chemical Week*, 1991b). Its main markets are the United States, Europe and Japan. Although Europe is among the largest markets, Case D cannot fully explore this market as will be explained below.

Inkco's group latex production (including Case D's) is about 350,000 tonnes per year. The group has productive units in Japan, the United Kingdom, Germany, Serbia, South Africa, the United States and Australia. Of this volume 40 per cent is produced by units within Cheminc and Inkco themselves, while 6–7 per cent is produced by each of the Serbian, South African and Australian units. Another 20 per cent is manufactured by the German arm, with the same percentage also being produced in the United Kingdom. As a consequence it is understandable that a solution of common interest should be found regarding sharing the European market. Obviously the German market (the largest by far) is almost exclusive to the German company. As a consequence Case D has to content itself with the British, most of the French, Spanish, Portuguese, Greek and the Scandinavian shares,[6] with local agents in all those regions and local stocks in several of them. For this reason 49 per cent of Case D's paper and board sales are directed to Europe, 38 per cent to the United Kingdom and 13 per cent to the Far East.

Case D's main disadvantage in the paper and board sector is its size. Paper mills are highly capital-intensive enterprises and need to develop a relationship with their suppliers whose closeness is proportional to their capital intensiveness. As described in one of Case D's publications:

> Before becoming a supplier to one of these companies, or even having a particular latex specification accepted for a new grade of paper by an existing customer, a painstaking process of two to three years' research and collaboration with the customer is often required. A high degree of trust must be developed for the relationship to be successful, and complete confidentiality is required from the supplier who may be supplying a number of closely competing mills.

Obviously it is, in principle, easier to have confidence in huge, well-known multinational companies than in small firms. Case D strives to guarantee its quality and reliability. By May of 1990 it had already achieved the accreditation of BS 5750 (part 2) and its European counterpart (ISO9002), ahead of some of its larger competitors. A few years earlier it had completed the implementation of a 'total quality management' programme. Another factor which is used as a competitive advantage against its larger competitors is, paradoxically, its actual smallness. This characteristic is fully explored in terms of responsiveness. According to information generated in the interviews, customers even have the home telephone numbers of Case D's employees, helping them to get a fast response to customers' orders or requirements at any time. In addition Case D seems to operate and fully understand just-in-time techniques.

The third segment where Case D operates is the so-called 'specialities'. Although smaller, it is the most complex in terms of market as it deals with applications as diverse as mortar additives and gloves manufacture. It tends to be higher in terms of margin and lower in terms of volume. Sales of specialities have kept their participation just above 10 per cent of the firm's turnover during the last five years. Specialities sales are shared among four main types of applications: building materials (33 per cent), non-wovens (33 per cent), dipping-products – where almost all of its nitrile latices are sold – (12 per cent) and adhesives (10 per cent). Adhesives is the application responsible for the smallest 'slice of Case D's market', but it is the one with the highest growth.

In contrast with all of its main competitors, Case D doesn't produce any of its own raw materials (that is, it is not vertically integrated). As observed above, such raw materials are styrene, butadiene and acrylonitrile, all of which are petrochemical commodities. As described in previous chapters the petrochemical industry has suffered heavy losses since the beginning of its present down-turn. Such a depressed economic environment is not, however, hitting all the sectors of this industry with equal force. Speciality chemicals do not present such a sharp cycle of 'booms' and recessions as do the commodities. Case D is a typical case of a company which is capitalizing upon the falling prices of commodities. As a consequence the company has been extremely profitable. During the 1992 fiscal year, for example, its profits reached almost 10 per cent of its giant parent Inkco's profits (a remarkable achievement considering that its turnover represents less then 2.5 per cent of the whole group). An easy explanation for such profitability can be taken from those raw-material prices. Butadiene, for example, had its contract prices fall from approximately US$870/tonne in July 1990 to an average of nearly US$420/tonne during fiscal year 1992 (see Figure 7.3). Within the same period styrene fell from DM1,730 to 850/tonne. Acrylonitrile kept more or less the same prices during the period, but between the middle of 1989 and 1992 it fell from DM1,715 to 1,280/tonne.

While Case D can take advantage of those falling prices, the same cannot be said about its large competitors (BASF, Bayer, Dow, Enichem) once they have integrated processes including those raw materials. Such integration may be an important advantage during periods of expansion of the industry as they can, in principle, obtain their raw materials in advantageous conditions. During a recession, however, there is a strong necessity to share costs or even losses throughout the whole supply chain, thus penalizing the business down-stream.

8.7 Case D's technology strategy analysis

8.7.1 Macro-level

From analysis at the macro-level, it is evident that since it was created, the synthetic rubber industry (which includes synthetic latices) had a clear trend towards the substitution of natural rubber in all its fields of use. This substitution process, however, seems to have stabilized one or two decades ago. Since then natural and synthetic rubbers are not competitors any more, but complement each other's markets (see for example *Elastomerics*, 1992). If that substitution process flattened, others are still rising. In the specialities market, SBR suffers competition from acrylic polymers and polyurethane in their adhesives' and non-wovens' segments (as well as the solvent glues). In building materials, acrylics and polyurethanes are followed by epoxy resins as the main substitutes for latices of SBR. In the paper and board area, acrylic polymers are, again, the main substitute. In textiles, PVC and natural latices are also important competitors.

Although it is quite difficult to see any common trend among the different companies competing for this market of synthetic latices, it seems that cost reduction is an obvious one as SBR latices are still expensive products with prices usually in the range of UK£1,000 to 1,500 per dry tonne.[7] In terms of technology one possible trend is the pursuit of an efficient technology for latex hydrogenation. Such a technology was already developed and successfully introduced to the market by some producers of dry SBR as well as dry nitrile elastomers. This breakthrough provided a considerable improvement in those products in relation to elasticity, heat, ozone and chemical resistances as well as cold flexibility (one of the main limitations on nitrile-rubber applications). Such success in the dry-rubber sector provides justified hopes of success also in latices.

8.7.2 Meso-level

The only technological innovations that a company of Case D's size can afford are shorter term and those with the most certain returns such as process development and end-use development. Such effort cannot be interpreted, however, merely as a kind of 'industrial fate'; it can be defended as an issue of strategic analysis as follows:

(a) Process development
  SBR latices can be block or random polymers. Among other factors, the process and the rate of monomers' addition influence a tendency towards the formation of blocks. In general it can be said that the monomer addition rate must be lower than that of polymerization in order to avoid storing unreacted monomer in the mixture which could favour long-chain block formation. If considering only this aspect, the advantages of a continuous addition of monomers are already evident in terms of product uniformity and quality. According to the interviews, however, the computerized continuous addition system installed in the middle 1980s provided Case D not only with improved quality but also some significant gains in productivity besides some novel blends.
(b) End-use development
  Basic technology for SB and NB latices is not difficult to buy or even to develop. Therefore, raw-materials suppliers, for example, can become a threat to traditional producers such as Case D. This threat, however, is not easily carried out: at the moment all chemical commodities-producing businesses are facing nil or negative margins preventing new down-stream investment. On the other hand when the industry starts its up-turn the prices of commodity raw materials can climb in a quite dramatic way, removing most of the attractiveness of the sector. Following this line of thought a successful investment in the sector depends on a clear view of timing for investment (and of its coincidence with the necessary financial availabilities). The effect of such

seasonal attractiveness collides with the continual investment in product 'end-use development' (that characterizes Case D) which, although
not necessarily involving high-tech, high-cost or even high-risk solutions, requires a long relationship with customers, entrenching the
traditional producer's position.

The two aspects discussed above may indicate a pattern (in terms of
time-span, risk and technological content for example), meeting at least
partially the definition of technology strategy adopted in Chapter 2.

Although the pattern described above doesn't seem to be changing, the
technology-strategy approach (as defined in Chapter 2) has strong signs of
change, as revealed by the fact that the firm is currently conducting a
project which aims to have 20 per cent of Case D's production based upon
non-SBR polymers within the next four years.

8.7.3 Micro-level

One interesting aspect of its technology strategy generation is the fact that
Case D exchanges technical knowledge with its parent companies Cheminc
and Inkco. Such exchanges are made mainly by the swapping of technical
reports and through group seminars and panels. Case D used to take the
initiative in promoting seminars concerned with paper coating (where
Case D is supposed to have some expertise over its parent companies) every
other year. Such a process can be seen as a part of the technology strategy
generation process.

Case D's technology strategy towards technical services and market-
orientated R & D, although clear and apparently effective, is not formal.
From financial goals set by its board, it is entirely up to the company to set
up its own goals and objectives. Although not confirmed by the company, it
can be expected that the technical-services' functional area receives much
more financial support than the R & D department itself as it has three
times more staff as well as its own laboratory facilities. Most of the R & D
projects are a consequence of technical services and other business areas'
requests, compounding a really 'market-driven' research.

As observed in the interviews, the research department itself uses one
tool for project selection and prioritizing. It is called a 'VAST chart'. This
chart involves three different sections all based on costs and potential
yields, leading to an evaluation of pay-back periods. According to responses
from the interviews, it can be said that the result of such evaluations (of
pay-back periods) is used for project selection, although it is just a part of
the decision-making process, which is mainly driven by what is generally
recognized as strategic to the company (which is made easier by its small
size). 'At the moment, for example, it is clear that the priorities are in the
area of paper coating and pressure sensitive adhesives, which present the
highest potential growth.' Also corroborating the definition of R & D
objectives and priorities is the participation of an external consultancy
which acts as an overview of technical tendencies.

8.8 Conclusion

Case D, although a limited company, is very small, even smaller than a typical SBU in the chemical industry. Little effect from corporate strategies was found in Case D as might be expected from a 'business controlled by three different corporations' (the corporate strategies and objectives may be diverse or even opposites nullifying each other). Therefore, most of the conclusions to be taken from this case-study concern only Case D itself and not so much its shareholder corporations.

As has been discussed earlier in this chapter, synthetic latices are not commodities or even fine-chemicals, but have the clear characteristics of specialities. As can be seen in Section 8.6, a high level of confidence between the latex supplier and the paper mill is absolutely essential for any business to be carried on between them. Such confidence can only emerge from dedicated technical assistance. Reinforcing such an observation, it is interesting to remember the disadvantages that Case D must overcome (mainly in terms of its relative size in comparison with its main competitors). All these factors, together with the informality (revealed for example by the availability of access for customers to call Case D's staff even at home), helps to define its business generic strategy as 'differentiation'. Even in its 'specialities' segment of the market, where capital intensity is much smaller than in the paper industry, Case D seems to carry a differentiation strategy which can be deduced by the fact that most of their clients in this segment buy exclusively from Case D.

The technology-strategy macro-level analysis failed to show any technological 'natural trajectory' after the fading away of the natural-rubber substitution process. The meso-level appraisal revealed a clear traditional approach (as defined in Chapter 2). All of its technological effort is put towards the strengthening of its present markets as no investment was aimed at new applications for SBR and NBR (the new project which aims to make 20 per cent of their production change into non-SBR polymers by 1996 may indicate that this approach will change). The present strategy is very much business-driven and seems to be paying off, but doubts remain about the company's perspectives in an eventual commodity up-turn.

The micro-level analysis shows a spontaneous, although effective, process of strategy generation without any methodology or formal approach. Its effectiveness may be associated with good management, corroborated by the short lines of communication peculiar to small companies. This type of approach will be referred to as 'informal'.

Owing to the limitations which Case D's size impose on its research funding and objectives, the only innovations that can be expected to be internally developed are those which Abernathy and Clark (1985) called 'niche creation' and 'regular'.

The appraisal of the five competitive forces shows a relatively weak situation in relation to powerful suppliers. No R & D response, however, can be taken to this threat as the levels of investment (as well as the attractiveness of those sectors) are prohibitive to a company of Case D's

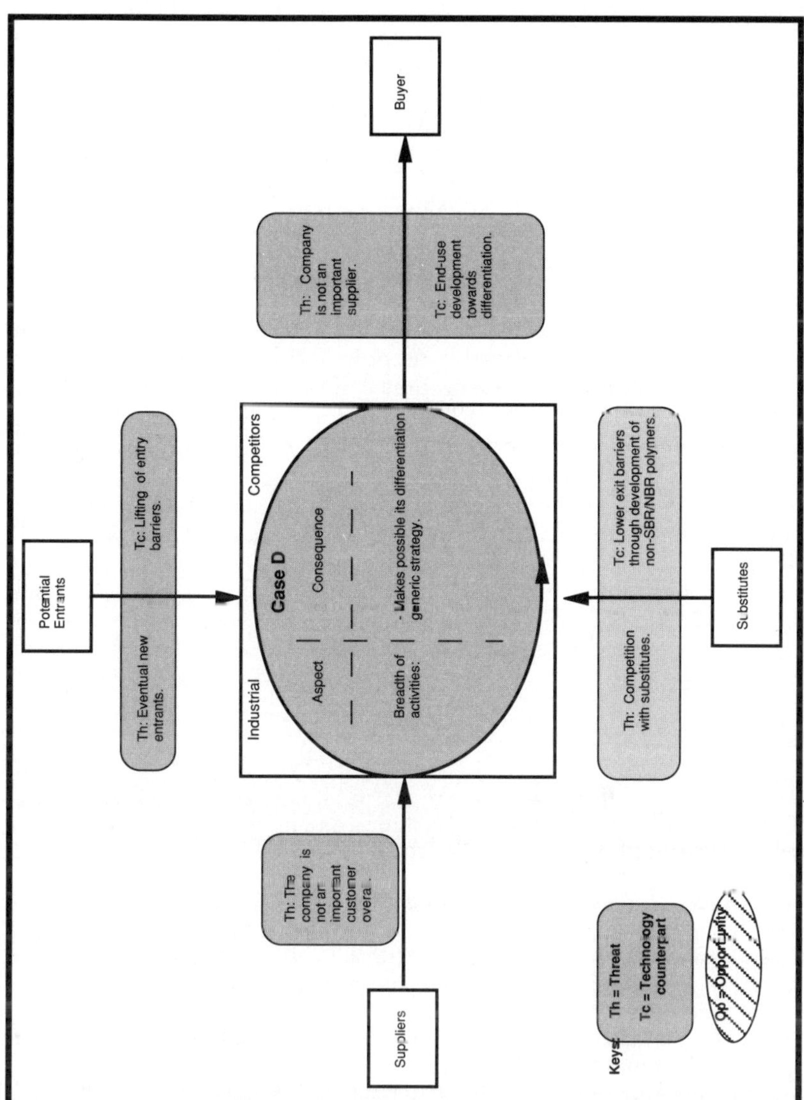

Figure 8.3 Technology and the five competitive forces (Case D).

size. It is clear that Case D's present profitability may be deeply affected by an up-turn in the commodity sector, as its larger competitors could count on favourable supplies of raw materials. Such a threat, however, is not so strong since there are several producers of styrene, butadiene and acrylonitrile, making it possible to look for alternative supplies.

Case D's R & D (up to the completion of this case-study) seemed entirely dedicated to its differentiation from its competitors through end-use development. It characterizes a counterpart (labelled B at the right side of Figure 3.2) to its small size and the bargaining power of its buyers.

Turning now to the new participants' dimension (Figure 3.3), there is a significant, although steady, threat of new entrants in the Far East (which is not yet a major market for paper coating). Nevertheless, the entry barriers to its markets (and particularly in paper coating and specialities) are relatively high in terms of end-use technology (entry barrier number 9 in Figure 3.3). Most of Case D's R & D results in a continuous raising of such barriers.

The risk of substitutes is also important as can be deduced from the penetration of acrylic and polyurethane polymers in its speciality markets. Here, the role played by their researchers can be seen in the development of new products (non-SBR/NBR). These comments about the relationship between technology and the five competitive forces are illustrated in Figure 8.3.

Case D's main threat then, seems to come from its industrial competitors and particularly from within its own group. As discussed before, Case D has a sister company of the same size in Germany with identical technology and markets. Such similarities make it extremely probable that both companies have to compete for any capital investment from their common parent Cheminc.

Notes

1. Cheminc, on the other hand, is run as a subsidiary company in relation to Inkco. Several of its businesses share technical and market information with the parent company; its organizational structure is, however, completely independent.
2. Cheminc has had major interests in Germany (even before its acquisition by Inkco); such interests have grown ever since. Among offices, plants and storage locations, Inkco has three different sites (including those of subsidiaries) in the United Kingdom, against nine in Germany. Among the German sites it has its corporate European Research Centre.
3. In most of its applications, natural rubber is further modified by vulcanization (a process for molecule cross-linking).
4. Although data are not available for all the different regions, it can be said that the fastest growth in terms of synthetic elastomers is among the 'thermoplastic polymer elastomers (TPE)'. This class of elastomers is expected to rise by 7 per cent per year in North America and 6 per cent in Japan (*Chemical Week*, 1991b).
5. A further variable that must be discussed in this brief review is concerned with the distribution of monomers throughout the polymer chain. An SBR latex will

have a chain of styrene (S) and butadiene (B) molecules connected to each other. The order in which those molecules can be combined can follow any binary combination within its range of molecular weight. A useful oversimplification is to classify such polymers as block polymers (when several molecules of the same monomer are connected to each other, for example: SSSSBBBBBBBSSSSSS) or random polymers when there are not such repetitive blocks.

6. With the sanctions to which Serbia was subjected in 1992 (as a result of the war in the former Yugoslavia), it can be expected that its unit will be facing great problems to keep producing. It is believed that the German company will capitalize upon such an incidence due to its geographical position.
7. Synthetic latices usually have about 50 per cent of their weight made of water. Their prices, however, used to be quoted on a dry basis (disregarding that water content). Curiously, their compounds are usually quoted in wet tonnes.

9 Comparative case-study analysis

9.1 Introduction

As a considerable amount of analysis has already been made and several conclusions drawn within the individual case-studies (Chapters 5 to 8), this chapter is intended to focus upon comparisons between the case-studies' analysis and some conclusions which can be drawn from this analysis.

9.2 General comparisons

This section provides some comparisons among the four different case-studies. The comparisons will be conducted at the corporate as well as at the business levels.

9.2.1 Inter-company comparisons

The case-studies conducted in this research took place in four different companies covering the complete size range found in the chemical industry. While Oilcomp and Chemicomp are very large companies with 115,000 and 130,000 employees respectively, Case D is a small enterprise with only about 200 employees, leaving Fibrecomp in the middle with its 23,000 employees. Table 9.1 shows some generic comparisons among the different corporations involved in this research. A first and important difference among the companies is that while the three largest are public companies, the last is a limited enterprise owned by two different groups. Among these two groups one is also a limited company fully owned by a third group (Inkco). As a consequence of its size and its managerial structure it could be observed that Case D, as a company, is comparable with the other three cases as SBUs. In other words, its size is comparable with Cases A, B and C.

Table 9.1   General comparison between the corporations studied.

	Turnover (UK£million)	Number of employees	Investment in R&D (% turnover)	Number of employees in R&D	Type of R&D (centrsd* descentrsd, buss* corp.)	Formal links between tech. and corporate strategy
Oilcomp	33,000	120,000	330 (1%)	4,000	Both, mainly business funded/ oriented	Director on the main board
Fibrecomp	1,900	24,000	40 (2%)	2,000	Centralised Business funded/ oriented	Director on the main board
Chemicomp	13,000	130,000	700 (5%)	13,000	De-centralised Business funded/ oriented	Exclusively dedicated director on the main board
Inkco	2,400*	NA	110 (5%)	1,500	Both NA	NA
Chemic	600*	NA	NA	NA	Centralised NA	NA
Britco	250	2,800	1.1 (0.5%)	NA	NA	NA

*UK£1=US$1.67

Also comparable is its dependence upon the parent group(s). In certain aspects this dependence seems less than that existing in the other cases. For example, it was revealed during the interviews that the only type of policy coming from the parent companies is financial. This indicates a larger degree of management freedom since big companies such as Oilcomp and Chemicomp certainly have policies for each of the functional areas within their groups. On the other hand Case D's size (see turnovers for example) is much smaller than those of the other SBUs under analysis. It is certainly true that its actual freedom in terms of investment, for example, is even smaller than that of the other companies studied. Consequently, Case D, although a limited company, will be analysed and compared with the other business units and its parent companies compared with the other groups (Oilcomp, Fibrecomp and Chemicomp) studied.

Looking to the data on relative investment in R & D (proportion of turnover), glaring differences can be observed. While Chemicomp and Inkco invest 5 per cent (or more) of their turnover in R & D, Britco doesn't even reach the 0.5 per cent level. Such a gap not only exemplifies but also explains the different nature of their patterns of investment. Chemicomp and Inkco are innovative companies whose driving force (as described by Tregoe and Zimmerman, 1980) is technology. Britco is a conglomerate whose driving force, based upon its financial expertise, has been return/ profit (see Section 1.4.3). Oilcomp's relative investment, on the other hand, is more difficult to comprehend if analysed separately. Its annual investment of 1 per cent of turnover can lead to the conclusion that it is not very concerned with research. However, considering its huge turnover (almost three times larger than Chemicomp), it can be argued that its effort is still quite considerable, representing an investment almost nine times larger than that of Fibrecomp.

Another observation to be made about corporate investment in R & D is that Fibrecomp is apparently underinvesting in technology. This impression comes, for example, following a comparison with Chemicomp which is a chemical company with a full range of products in its portfolio. Despite the large participation of petrochemicals and other heavy bulk-chemical commodities (whose products are not as research-intensive) in its portfolio, it still invests more than 5 per cent of its turnover on research. Meanwhile, Fibrecomp, which defines itself as a materials company (a more research-intensive sector), does not go beyond a modest 2 per cent. (This subject is further discussed in Section 9.5.)

Another interesting observation is that there seems to be a link between the relative investment in R & D and representation of the R & D function on the main board. Chemicomp, which has the largest investment in research (closely followed by Inkco) is the only company having a representative dedicated exclusively to the R & D function on its board.

The most striking observations from the inter-company comparisons, however, are those shown in the column concerned with 'type of research and development' in Table 9.1. In that column companies' research facilities are classified as centralized when they are combined under any 'overall corporate interest' such as that used by Fibrecomp (described in Chapter

6) or as decentralized, when facilities are spread through different sites. A parallel classification summarized in that column is concerned with research funding. Here the research efforts of the different companies are classified as business or corporate funded. Although these two classifications (funding and centralization) would seem directly connected, this is not the case. Fibrecomp, for example, has centralized, although business-funded research facilities. The combination of the two classifications seems to provide a good idea about how far these companies are concerned with the linkage between business strategies and technology. As can be observed in Table 9.1 all companies, with the exception of Chemicomp, have a balance between concentration and corporate-funded/orientated R & D and division and business-funded/orientated research. The only extreme seems to be Chemicomp which, in spite of the huge size of its effort, has almost only business-orientated technological concerns. This observation will be further analysed in sub-section 9.2.2 and Section 9.5. Additional conclusions about corporate technology strategy also will be presented in Sections 9.3 and 9.4.

9.2.2 Inter-businesses comparisons

As described in Chapter 7, considerable difficulties were experienced in analysing Case C as a single business unit (because of the diverse nature of its products and management). In order to allow a useful comparison to be made with the other case-studies, it has proved necessary (in several of the comparisons) to split that study into two different analytical parts, which will be called cases 'C1' (adiponitrile/HMDA) and 'C2' (nylon polymers and fibres).

As happened with the corporations, the four businesses analysed in this research also covered a wide range of sizes. From 8,000 employees (and a UK£700 million turnover) in Case C to 200 (and UK£35 million) in Case D. Table 9.2 shows a categorized selection of data about these SBUs. Although there is no direct link between the size of the corporations and their SBUs, in the four case-studies conducted here, the SBUs' sizes were roughly proportional to the corporation sizes.

With the exception of Case D, all the businesses have turnovers of the same order of magnitude, in addition the range of investment in R & D is narrower than occurs in the corporations. Looking at the number of employees in R & D is also an informative exercise. Case A is considerably more intensive in this aspect. There are, however, two points to be considered. The first is that they are the only case conducting a major innovation (see Sections 4.4.1 and 5.7.2). The second point is the fact that the research and development of this innovation (propane ammoxidation) has been conducted in the corporate research facilities. This second point explains how this business can allocate a workforce as large (at least) as 25 per cent of its whole work-force. Oilcomp-published brochures on chemicals and particularly on research give the impression that most of this work-force is corporate work-force allocated to this task.[1] An analysis of the

Table 9.2 General comparison between the different case-studies.

	Turnover (UK£ million)	Number of employees	Investment in R&D (% turnover)	Number of employees in R&D	Type of product	Formal links between tech. and corporate strategy	Typical price US$/t	Approximate Capacities 1000 t/y (%world)
Case A	250	400	8.7 * (3.5%)	More than one hundred	Commodity	Business development manager	700 to 800	630 (15%)
Case B	200	NA	3.5 * (1.5%)	35	Speciality	R&T organisational and budgetary structure	2000 to 3000	150 (6%/60%)**
Case C C1	700	8000	3.5 (0.5%)	200	Pseudo-commodity	Business development manager	2200 to 2600	90 (9%)
Case C C2			10.5 (1.5%)		Speciality		2000 to 6000	200 (6%)
Case D	35	200	0.5 (1.5%)	7	Speciality	(Informal)	1800 to 2200	70 (4%)

* – Author's estimate

** – Total/Dyed fibres

proportions shows that Case D, despite the small absolute number of researchers, has a considerable effort in R & D (the percentage of its staff involved in R & D is larger than that of Case C).

In terms of type of product there are important observations to be made. None of the products analysed here is typical of the group where it was placed (the groups are commodities, pseudo-commodities, etc.). The only exception is perhaps acrylonitrile, which is a typical petrochemical commodity, but nevertheless has some outstanding particularities such as Case A's technology monopoly). Hexamethylenediamine (HMDA) could be considered a petrochemical commodity as well, although its easy decomposition makes its international trading a rare occurrence. Most often it is commercialized as adiponitrile (ADN), which is further hydrogenated to HMDA near its consumption site. Even ADN is not a typical commodity since there is just a handful of producers around the world (as a consequence it is not easy to buy its technology – a quite unusual circumstance for commodities in general). In terms of nylon and acrylic fibres there is an enormous variety of grades (a few dozens for acrylics and a few hundreds for nylon). This factor demonstrates the virtual impossibility of a precise classification. According to a consultancy company specialized in nylon and nylon fibres which participated in this research, most of the synthetic fibres were commodities priced at speciality prices up to 20 or so years ago. Presently (according to those consultants), they are usually specialities priced as commodities!

As can be understood from the above discussion, the classification of chemicals may be controversial and very much dependent on the chosen criteria. As the main concern of this book is the management of technology, it seems quite reasonable to classify those products from this point of view. As the innovation process in the fibres sub-sector is very much concerned with product innovation (in sharp contrast with what happens in true commodities), they are here generally classified as specialities. The synthetic latices classification is easier to make as it is an 'almost' typical speciality (the word 'almost' is used here because large amounts of its basic polymers are produced throughout the world).

In terms of typical prices the numbers presented in Table 9.2 are also very generic and imprecise as there is a myriad of variables involved when setting up their prices. Some of the most important are:

– Contract versus spot market prices;
– Type of contract (FOB, C and F, CIF etc.);[2]
– Location (main areas are: West Europe, Japan, Asia/Pacific, North America etc.);
– Grade (fundamental for fibres);
– Volume to be traded.

The numbers presented in the table are intended to provide only a rough idea of their relative positions in Europe at the end of 1992.

In terms of capacities, acrylonitrile has, by far, the largest global production (which could be expected as it is the main 'building block' for most of the products included in this research, see Chapter 3). Among the four

SBUs the most important globally is Case A as it has a bit more than 15 per cent of the world capacity, and 100 per cent of the world production is now based on its fluid-bed propylene ammoxidation technology. Case B, while responsible for only about 6 per cent of the world production of acrylic fibres, may represent more than ten times this percentage in terms of dyed-acrylic fibre. Case C, with 6 per cent of the world production share, is much more important in terms of the European market, where its share reached some 40 per cent. Case D has, again, the smaller numbers with only about 4 per cent of the world production.

9.3 Technology strategy

A first overall conclusion is the fact that all the managers interviewed (particularly those in charge of business-development functions) showed familiarity with the concept of technology strategy as defined in Chapter 2. Different approaches were found in the promotion of this linkage, and the conclusions about the types and consequences of such approaches will be discussed in the sub-sections below.

9.3.1 The technology strategy, its genesis, models and tools

Among the models described in Chapter 2, those derived from the work of Porter (1985) and Wilkinson (1987) (see Figures 3.2, 3.3 and 3.4) were the most comprehensive. All the R & D counterparts to strategic threats or opportunities observed in the four case-studies can easily fit into those models. The generic and abstract character of those models, however, limits their utility for those outside the industry. Other techniques such as technology audit and technology versus business-quality matrices were found to be in use. Although not providing a straight answer to the strategic problems and questions, these more sophisticated models may be useful in strategic analysis as will be described below.

Another conclusion to be drawn in terms of technology-strategy management in the four case-studies is that the formality and 'sophistication' of management is proportional to corporate and business size (and complexity). The extremes found in this research were Cases C and D. Case C is not only the most complex business in this analysis but also the corporation of which it is a part is the most complex among the whole group (the complexity can be easily recognized when looking at the diversity of products and geographical areas where the different companies and businesses are operating). Case C's management can also be easily identified as the most 'sophisticated' among the whole group. It was the only company and business where specific techniques and tools for management of technology are widely used. Case C also employed a business development manager with expertise on research and marketing of its products. This manager was formally in charge of linking those two functional areas.

The other extreme of our comparison (Case D) not only doesn't use techniques specific for the development of technology strategy but also did not have any formal approach to the subject. Like the observation made above about sophistication versus quality, it is important to note that Case D's informality is not synonymous with low quality of management.

In the other two case-studies the approach to technology-strategy management could be classified as an intermediate between the two extremes above. Both have a formal approach, Case A also has its business development managers whose duties are very much like those of Case C. The 'business development manager' interviewed, however, wasn't aware of any managerial tool designed specifically for that job.

Case B, although not having a manager exclusively in charge of such matters, has a relatively sophisticated budgetary and organizational structure in its 'Research and Technology Department'. This managerial structure (described in Section 6.3) keeps R & D formally attached to the individual businesses while also keeping a balance between business and corporate-orientated research.

9.3.2 Natural trajectories, common trends and strategies

An important result evolving from what Mulder and Vergragt (1990) called 'technology strategy macro-level analysis' is that while 'natural trajectories' towards processes employing cheaper raw materials were easily identifiable in the case-studies related to commodity and pseudo-commodity chemicals, the same could not be said about the specialities. For this third group of chemicals the only common aspect observed in their technology strategy is a search for higher specialization, thus creating a paradox since their only common characteristic is searching for unique products and markets.

A further important conclusion to be drawn from this level of analysis is that among the innovations identified in this study all those concerned with commodity and pseudo-commodity products were process innovations. In the group of speciality producers, the innovations identified involved both process and product innovations.

Among process innovations the only potential major innovations (as defined by Stobaugh, 1988) are the new ammoxidation process under development in Case A and the catalytic acrylonitrile dimerization process developed by Case C1. Cases C2 (nylon fibres) and D (SBR and NBR latices), which are more 'speciality-like' producers although also promoting minor innovations in terms of process, have innovated mainly in terms of product. Within this group, Case D's product development is particularly interesting as, albeit quite simple in comparison with the other cases presented here, it is extremely relevant to its business strategy as discussed in the next sub-sections.

A last aspect to be emphasized from this point of view is the location of all the different innovations observed in the research on Abernathy and Clark's (1985) transilience map. The conclusions to be drawn from this chart are more academic in nature but can also be useful when first analysing an industrial technology. Figure 9.1 shows the position of the main innovations identified in each of the case-studies. This figure also shows their classification as 'true innovations' (as the concept is used in influential works such as the SAPPHO project) or as an 'internal innovation', as discussed in Chapter 3. In Figure 9.1 they are also classified as major or minor innovations as defined by Stobaugh (1988). Three conclusions can be drawn from that figure: in general the innovations (of any type) for each individual product follow an anti-clockwise move from its first architectural innovation. Second, the internal innovations are always minor ones, while the 'absolute' or 'true' innovations may be either (minor or major). It can also be seen that major innovations are always architectural or revolutionary (following Abernathy and Clark's definition), while minor ones usually are associated with the concepts of 'niche creation' or 'entrenching'.

9.3.3 The interdependence between corporate and technology strategies

Although identifying much stronger links between business and technology strategies, some evidence of a link between corporate and technology strategies was also identified. A convenient starting point for this discussion is provided by the functional link between the two areas under analysis (corporate strategy and technology). All the corporations studied have a technology director with a seat on the main board[3] but only in Chemicomp is this director exclusively responsible for R & D. In the other cases the director of R & D or technology is also responsible for other functional areas, diluting his attention and revealing the relative importance of this matter in the different corporations.

Other important sources of information are the corporative goals. Oilcomp's main goal is to reduce its financial problems caused by its high debt burden. In order to do that, it is successively divesting businesses not directly related to its traditional areas of expertise (oil and chemicals). Fibrecomp's main goal in the last few years was to move its portfolio from commodity and cyclical businesses (up to a few years ago heavily based on the British textile industry) towards products of higher added value. This move has been basically completed and was accomplished through a strategy of diversification around its core technologies (surface, colour, polymers and materials). Chemicomp has also been concerned with moving away from commodities to high-value, high-tech businesses. In contrast with Fibrecomp, however, its strategy has been to invest only in businesses that can be globally competitive (implying a high share of the world capacities and markets).

In Chapter 2, the firms' approaches to technology strategies were divided into two broad categories: traditional (when basically concerned with existing products and markets) and competence-based (when mainly concerned with the development of new products/markets). The evidence found in the case-studies, however, is that no approach is purely traditional or competence-based but rather a mix of both. Nevertheless, it was still possible to identify a large component of one or other of those theoretical

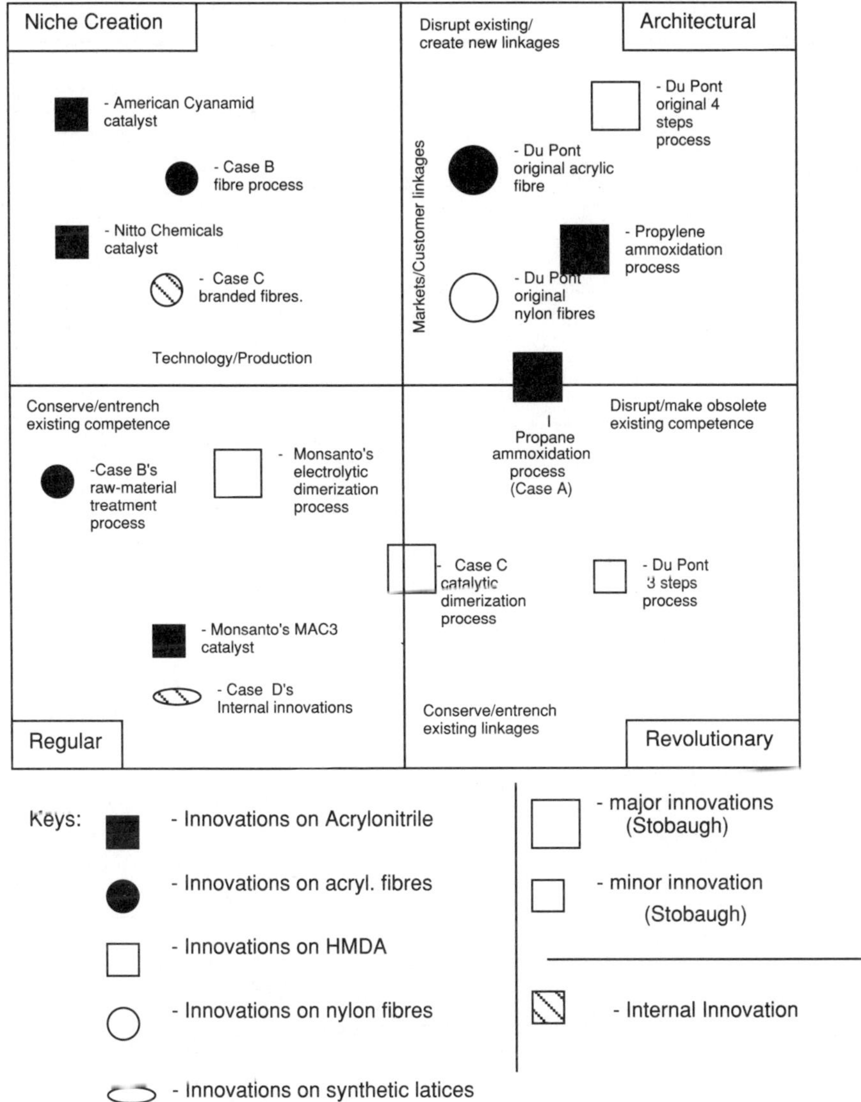

Figure 9.1 General positioning of the different innovations identified among the case-studies in 'Abernathy and Clark's transilience map.

approaches. Chemicomp and Inkco were easily identifiable as having a more competence-based approach, mainly characterized by their con-tinuous move towards new products and markets. Although such a move towards new products and markets could also be found in Fibrecomp (eventually and in a smaller degree also in Oilcomp) its movements were

Table 9.3 General conclusions about the corporations.

	Corporate Goal	Corporate Strategy	Technology strategy approach	Use of technology as input on corp. strategy
Oilcomp	Debt burden reduction	Divestment of non-core businesses	Mixed	Apparent
Fibrecomp	Higher added value product portfolio	Diversification around core technologies	Traditional	Apparent
Chemicomp	Higher added value product portfolio	Invest only in globally competitive businesses	Mixed	Clear
Inkco	NA	NA	Competence based	Clear
Cheminc	Turnover > US$ 2 billion by year 2000	NA	NA	NA
Britco	NA	NA	Traditional	NO

almost always made through acquisition, while Chemicomp and Inkco depended more on internally developed technology. These observations are summarized in Table 9.3.

When comparing Tables 9.1 and 9.3 a consistent and coherent matching of information can be seen around three groups. The first contains only Britco which stands out not only for its reduced investment in R&D but also because of its traditional approach to technology strategy with a result of an unidentifiable use of technology as an input to corporate strategy. The intermediate group includes Oilcomp and Fibrecomp with an intermediate level of investment (relative to turnover) in R & D and a mixed or traditional approach to technology strategy. Both companies also have a R & D director (not exclusive) on the main board. In these two case-studies the use of technology as an input to corporate strategy is apparent, although not as evident as it is in the next group. The last group is composed of Inkco and Chemicomp, both with an outstanding level of investment and, at least in Chemicomp's case, with a director exclusively responsible for R & D who sits on the main board. Both firms also have a more competence-based approach to their technology strategies. It can be said that in this last group technology is clearly seen as an important input to corporate strategy.

Although important, all the parameters used to classify Cheminc and Chemicomp in the leading group discussed above are not enough to ensure proper management of technology, as will be demonstrated below.

9.3.1 The interdependence between business and technology strategies

Before discussing the aspects related to the link between technology and business strategies, it is important to review the businesses' competitive position and strategy in each of the case-studies. The two taxonomies employed here were: Porter's generic strategies and the classical BCG's (Boston Consulting Group) matrix. All the managers interviewed discussed both taxonomies with relative ease. In two cases the business classification as a 'cash cow' was not just a conclusion from the analysis but also the managers' view about their own SBUs.

Cases A and B can be easily classified as 'cash cows'. Both of them have a large share of the world market[4] but are in an industry of low attractiveness (whose growth is expected to be only vegetative). Case C1 had the worst position in the group. Not only is their competitiveness very low but also the market is not attractive; consequently it was classified as a 'dog' (see Figure 1.6). Case C2 presents higher competitiveness than C1, coming into the 'cash cow' classification. Finally Case D has better competitiveness (at the moment) and is in a slightly more dynamic industry (in terms of growth). Both parameters, however, are not good enough to put this business in a different quarter of the matrix. Most important is that an eventual upturn in the commodity market may be enough to make it move to a 'dog' position.

In terms of Porter's classification all the case-studies, with the exception of Case B, showed a strategy towards differentiation. Case B's generic strategy seems better classified as 'differentiation-focus', as the business is clearly aiming at the dyed-fibres niche where it, in spite of its lower costs, has a differentiated position. Table 9.4 illustrates some of the conclusions and observations found in the different case-studies which will be discussed below.

In terms of the approaches to technology strategy at the business level (traditional or competence-based), the SBUs seem to follow the same approaches as their parent companies. Case A, although investing heavily in its main product (acrylonitrile), has also invested in new technologies and markets such as its maleic anhydride process, resulting in an approach that could be classified as mixed. Case B has invested exclusively in acrylic fibres characterizing a traditional approach. Case C, although only investing in nylon fibres, has launched several new products (or at least new brands) over the last few years, opening new markets and resulting in a mixed approach. Finally, Case D has been clearly traditional but its announcement of a goal of 20 per cent of production based upon non-SBR/NBR latex for 1995 shows a move which apparently represents a mixed approach.

As can be observed by comparing these approaches with those of the parent companies, there is no sharp contrast. The only differences (not an incongruence by itself) are Case C's mixed approach and Chemicomp's competence-based one. In case D, on the other hand, a mixed approach combines with a balance between Inkco's competence-based and Britco's traditional approaches. This observation opens a way for a partial conclusion that, in principle, businesses' approach to technology strategy follows those adopted by their parent companies. The main conclusion about this analysis, however, is related to the main objective of this research which is an understanding of the relationship between business strategies and technology strategies. In accordance with what has been observed throughout the entire study, there is no direct connection between Porter's generic strategies and/or BCG's classification and the observed approaches to technology strategies. (The absence of a visible linkage in the four case-studies, however, doesn't mean that in the industry as a whole such linkage cannot be observed.)

A further conclusion common to all the case-studies is well illustrated by Figure 3.3; most of the technological effort found in each case was clearly resulting in higher entry barriers for new entrants. Case D's new programme involving non-SBR polymers, on the other hand, is the clearest example of the use of R & D counterparts to lower its own exit barrier.

9.4 Conclusion

The three levels of analysis of technology strategy developed by Mulder and Vergragt proved to be a useful part of the methodology. It not only provided a framework for analysing the companies from a common point of view, but also provided some interesting conclusions about the individual technology strategies themselves. Its association with in-depth case-studies, however, didn't help in providing evidence of a link between the technology-strategy approach and business generic strategies and/or their position in the BCG's classical matrix of business competitiveness versus industry attractiveness. It may be suggested that owing to the relatively subjective nature of all the taxonomies involved (Porter's, BCG's and the

Table 9.4 Some of the main conclusions and observations from the different case-studies.

	Generic Business Strategy	BCG'S Classification	Technology Strategy Approach	Innovations observed in the case study	Main Innovation's Classification
Case A	Differentiation	Cash Cow	Mixed	Propylene/ Propane Ammoxidation	- Major - Absolute - Process
Case B	Differentiation /focus	Cash Cow	Traditional	Dyeing process	- Minor - Absolute - Process
Case C — C1 / C2	Differentiation	Dog — — — Cash Cow	Mixed	Catalytic dimerization process — — — Branded fibres	- Major - Absolute - Process — — — - Internal - Product
Case D	Differentiation	Cash Cow	Traditional ↓ Mixed	End - use development	- Internal - Product

Technology-Strategy Approach), only a wider study could provide some evidence. Further overall and main conclusions are drawn in the next chapter.

Notes

1. It is important to highlight the fact that Case A was the only company not to allow the author to visit its research installations.
2. FOB means Free On Board, and when placed beside a listed price means that the seller is responsible for placing the product on board the buyer's ship.

   C&F means Cost and Freight. In this case the seller is paying for freight and costs up to the named destination.

   CIF means Cost, Insurance and Freight and, in contrast with C&F, implies that the seller is paying for insurance. Other usual types of quotations are *Delivered Sale*, when all risks and costs go to the seller up to the permanent hose connection at the receiving site. *Ex Ship*, when the seller is responsible for everything until the ship's arrival at the discharge port.

3. Almost all the discussion about corporations will exclude Case D's parent companies (Inkco, Cheminc and Britco). There are several reasons for such exceptions. First, it must be considered that it has three parent companies, contrasting with only one in each of the other cases. This fact makes any corporate influence more 'diluted' than in the other cases. Cheminc is not a public company so there is much less information available about this company than about the others. Inkco, although a large public company, is not directly linked to Case D. As observed in this last case-study, since Case D is much more dependent on Cheminc than on Inkco, the availability of information about this latter company doesn't help to provide a better understanding. The situation is slightly worsened by the fact that both companies are based abroad. Britco, although in the United Kingdom and a public company, is easily defined as a financial conglomerate and consequently having little influence upon Case D's technological issues.

4. Case A has only 15 per cent of the world production but is responsible for more than 30 per cent of the merchant market. Case B has a comparable position with only 6 per cent of the whole production but more than 50 per cent of the dyed-fibre niche.

10 Conclusion

10.1 Introduction

This research generated three different types of conclusions:

(1) those related to each individual case-study;
(2) those which relate specifically to the inter-case study comparisons; and,
(3) those which are understood to be the main contributions of the research to the existing knowledge about the relationship between business and technology strategy in the chemical industry.

The first type of conclusions have already been presented in each of Chapters 5–8. The second category, which can be called 'overall conclusions', was described in Chapter 9, while the major and broader overall conclusions will be drawn in this chapter.

This concluding chapter is also intended to present a concise overview of the research. This overview starts in Section 10.2, which was structured to act both as a summary of the four theoretical chapters of this book as well as being a contribution to the alignment of the conclusions with the general perspective of the research.

Section 10.3 concentrates on what was observed in terms of the links between technology and business strategy. Section 10.4 presents the implications of the broader conclusions for the field of study. This section is divided into several sub-sections related to the evidence found in the case-studies in terms of the emergence of a new paradigm in the industry (Section 10.4.1). Among such evidence there are two groups of managerial practices. The first group, which consists primarily of commercial managerial practices is generically referred to as 'de-commoditizing' and is discussed mainly in sub-section 10.4.2. The second is referred to as 'strategic management of technology' and is primarily discussed in sub-section 10.4.3. Section 10.5 summarizes the conclusions, and Section 10.6 describes the possible implications of these conclusions.

10.2 Summary of research

10.2.1 From Chapter 1

The use of the concept of strategy as a business subject received its first thrust at the academic level in the mid-1950s (Drucker). In the marketplace the practice of strategic planning was initiated from the 1960s (General Electric). Since then it has passed through several phases and dozens of definitions were drawn within the three main schools of thought. All the definitions, however, are based upon just four ideas: 'importance', 'long term', 'pattern' and 'interactions with the business environment'. Over time the three schools of thought appear to have moved towards a holistic view in which the four ideas cited above are analysed along three dimensions: 'process', 'variables' and 'environment'. This holistic view was labelled 'strategic management'. It was also observed that there is no 'best' definition for strategy and, as a consequence, that the definitions to be used in the research should be chosen based upon its objectives and methodology. The 'natural' definition for the purposes of this research was then assumed to be:

> A pattern of resources and competencies deployed to achieve a goal or objective which involves environmental interactions.

10.2.2 From Chapter 2

Studies of the relationship between 'strategy' and 'technology' have increased during the last 15 years. There are three different loci of analysis of this relationship in terms of firms and enterprises:

- The corporate level, where the main concern is the search for technological synergies among the different businesses in a corporation's portfolio and where two generic types of approach were identified; the 'traditional' (based upon sustaining existing markets) and the 'competence-based' (which search for new markets based upon the existing core technologies).
- The individual business level, where the two approaches described above can also be identified, although technology's main role was identified as 'generation and maintenance of competitive advantages'.
- The functional level, where the definition of strategies for functions such as production, R & D, marketing and others are analysed.

Considering Chapter 1's conclusions, a definition for business strategy was assumed to be:

> A pattern of resources and competencies deployed through environmental interactions to achieve a business goal or objective.

Technology strategy was defined as:

A pattern of techno-scientific resources and competencies deployed to achieve a technological goal or objective defined from the analysis of the business's competitive forces.

10.2.3 From Chapter 3

A methodology based upon in-depth case-studies in a single building block of the petrochemical industry was selected as appropriate for this research. One supply chain was chosen and its general characteristics were described. The framework for technology strategy in each of the case-studies was defined, based upon the model built by Mulder and Vergragt (1990). Within each of the three 'analytical spaces' which make up the framework, some of the main models available in the literature were reviewed.

10.2.4 From Chapter 4

The chemical industry evolved through different phases and paradigms. The role of innovation within this evolution was always a major one. In the early years of the industry, innovations were mainly attributed to individual inventors (sometimes working in their backyards). Owing to its 'strategic' importance and complexity, innovation activity became more and more the domain of professional researchers. The growing scientific nature of the innovations in the industry led to the creation of huge research centres which, associated with the capital intensiveness, resulted in major innovative activity becoming a task almost exclusive to large corporations. In the modern petrochemical industry this exclusive character can be exemplified by the research in processes aimed at cheaper feedstock. Down-stream from this feedstock the industry is facing some new challenges as the large corporations try to move away from the highly competitive market of commodities towards chemical specialities. The challenges cited above seem to be associated with, among other factors, a new 'techno-economic-paradigm' whose emergence stimulates speculation about the industry's future.

10.3 The relationship between technology and business strategy

The first and most straightforward conclusion to be drawn from this analysis is that in all the case-studies the concern about the linkage between technology and business strategy is very evident. All the companies studied are clearly trying, in some way, to promote an effective link between functional areas concerned with technology, such as research and development, with those concerned with business strategy such as marketing, strategic planning and business development.

The analysis showed that there are several different approaches to promoting the linkage between technology and business strategies:

- The first approach was called 'personnel'. It was characterized by the existence of jobs whose description mainly included the formal linkage between technology and business strategies. Such an approach was identified in Case A and Case C.
- The second type of approach, which was found in Case B, involved an organizational and budgetary structure which imposes the sharing of medium- and long-term research projects among the individual businesses in accordance with their core technologies, while also keeping business and technology development in intimate contact. This model was called 'organizational'.
- The third type, labelled 'informal' in Chapter 4, was identified in Case D and characterized by the absence of any formal linkage between the two areas.

As was described in the previous chapter, there are different degrees of 'sophistication' in each approach. Case C was identified as being the most sophisticated, with Case D at the other extreme. There are two very important aspects to highlight about the 'sophistication' referred to above: the first is that in spite of being the company where such techniques and tools are most widely used (Case C), they were used much more as a 'common language generator' than as an effective decision-aid as described in Sections 7.7.3 and 7.8. The role played by such techniques is to put the different projects in the company under the same focus of analysis. Decision-makers in very different managerial cultures are forced to compare their projects and priorities using the same language and parameters. The second point is that the term 'sophistication' is not used here as a synonym or even as a correlated term to 'quality' (as discussed in Section 9.3.1).

The different approaches can also be characterized by the higher or lower degree in which technology development is linked to the business strategies. Although all companies analysed presented some mix between business and corporate driving, the most radical approach was found in Chemicomp, which was the only company (among the large ones) with almost exclusively business-driven research.

Probing each business as an individual unit of analysis, however, provided several other important (and specific) conclusions as listed below:

- Case A's new propane ammoxidation process will help it keep its position as the largest producer and can even enhance all the characteristics that differentiate Case A from its competitors. An eventual plant based upon the new technology and situated in the Far East will further spread its production around the world, enhancing its already strong position in terms of guarantee of supply and speed of response (if the company keeps its present prices, it may also provide a higher margin of profitability).
- Case B's present process technology for dyed fibres is so important that it not only makes possible but also justifies its business strategy to be classified as one of differentiation-focus. Its investment in process improvement (reported in Chapter 6) should maintain its leadership until

the emergence of a radical innovation.[1]
- Chemicomp's indecision about investing in Case C1's catalytic dimerization process may be the main reason for its low profitability (and further divestment as explained below). However, its investment in Case C2's product development (smaller than that required for the catalytic process) provided the main reason for its differentiation and the consequent extension of its 'life' within that corporation.
- Case D's technology strategy, although involving only minor or incremental innovations, retained its process competitiveness and also resulted in end-use development. This made possible its differentiation, which is helping to keep this company competitive among its much larger rivals.

The obvious conclusion to be made from the comments above is that technology has a fundamental role to play in the genesis and implementation of business strategies in the chemical industry. The strongest conclusion, however, may be made from the analysis of Case C as a whole. It is very important, however, to highlight a change which took place after the conclusion of the empirical work in case-study C. This was the divestment of 'Case C' by its parent company 'Chemicomp'. By April of 1992 Chemicomp and Du Pont announced that they intended to swap all Chemicomp's nylon business (Case C) for Du Pont's 'acrylic businesses' in America.

This case-study showed the real danger of an absence of a technology strategy (or pattern as defined in Chapter 1). The business invested heavily in the research and development of a catalytic dimerization process for production of adiponitrile from acrylonitrile (further contacts with an ex-chairman of that division revealed that the pilot plant alone cost some UK£5 million). This development was a potential major innovation. The research was successful but that pattern of investment was broken and an industrial plant was never built.[2] In short, the absence of a corporate technology strategy towards that business signified a waste of several million pounds on research and development or (worse still) of possibly dozens of millions of pounds on the divestment of a business that might have proved competitive in the future. This seems to be very much the view of some analysts. Fryer (1992) for example, says:

> While Du Pont is to be complimented on its acquisition, the sale must be seen as a bad decision on 'Chemicomp's' part. By the end of 1992, Du Pont will be a world force in both nylon 66 and polyester, while 'Chemicomp' will have abandoned both. It need not have been so.

The absence of a proper technology strategy towards nylon chemicals in Case C may be associated with what was termed by Achilladelis *et al.* (1990) as 'corporate technological tradition' (CTT). This work (cited in Chapter 4) describes how 'home-made' radical innovations used to promote a 'long-term compromise' in those innovative corporations with products/markets where the radical innovation was introduced. Although Chemicomp had been involved in some important innovations related to nylon chemicals and fibres, it did not produce any 'radical' innovation. It can be theorized that although nylon was part of Case C's culture, the same was not true for its parent company (Chemicomp). Such a conjecture seems to be reinforced by the absence of a pattern in Chemicomp's behaviour in relation to these products (as described throughout Chapter 7). In the light of this 'corporate technological tradition theory', the absence of such a 'pattern' can be seen as a continuous shock of cultures between Case C (a business with corporate dimensions) and its parent company, Chemicomp (a giant corporation).

As will be clearer at the end of this chapter, several of the conclusions resulting from the analysis of the relationship between technology and business strategies can be associated with the emergence of a new industrial paradigm. These conclusions, as well as the reasons for such association are described below.

10.4 New paradigm and managerial practices

10.4.1 The emergence of a new paradigm

Reviewing some of the discussion about the new industrial paradigm introduced in Chapter 4 and comparing its main issues with what has been observed in the different case-studies is an important exercise. It not only makes it possible to situate the chemical industry within the broader managerial trends observed in world industry as a whole but also makes it possible to ponder the possible influences of this new paradigm upon the chemical industry's pattern of innovation (also discussed in Chapter 4). In order to do that, some of the main features of the new paradigm discussed by Perez (1983) and Bessant (1991) (among many others) are listed below. Each feature is followed by observations from the individual case-studies:

1 – 'Greater emphasis on non-price factors'
   Various non-price factors observed had a major role in all case-studies. Among the case-studies, those whose business strategies are more dependent upon these factors are Cases C2, B and D. Nevertheless, even Case A, in spite of being a commodity producer, strives to differentiate itself from its competitors through factors related to the quality of its product and technical services.[3] Among several non-price factors one of the most prominent is the fact that all companies studied here implemented, during the last three years (or are now implementing), a total quality programme.
2 – 'Greater emphasis on flexibility in technology' (which may be complemented by Perez's observation on 'flexibility in process investment and in product mix rather than the dedicated capital equipment and standardised products range characteristic of the Fordist mass production system')
   Such 'targeting' on flexibility was a very clear issue from Case B's

process and strategy analysis. If Case A cannot have as flexible investments and plants, at least its portfolio of technologies will present some flexibility from the development of its propane ammoxidation process. It will, for example, be able to choose the most strategic raw material for future investments (a flexibility that no other producer will have). Also Case C2 shows a 'flexible mix of products', although Case D is the outstanding example of versatility in product mix and organizational structure, which is in line with the third feature of the new paradigm as visualized by the authors cited and discussed below.

3 – 'Greater emphasis on flexibility in organizational structure'
Case D's compounding division and manufacturing sites make up one of the clearest examples of organizational flexibility. Furthermore, Case B's research and technology division's organizational structure can be very flexible, easily absorbing eventual changes in its funding for example.[4]

4 – 'Changing relationships within and between organizations'
A common feature observed in all case-studies and very much concerned with 'changing relationships between organizations' is the 'just-in-time' system. The most evident example of such a change could be found in Case B. This firm's computers are connected to a major clothing retailer's computers, providing data about the selling trends of their fibres. This example stands out even more if we consider that there are several other manufacturing levels (yarn, fabric and clothes) in the chain between the two companies.

5 – 'Information-intensity rather than energy-intensity in products and processes'
Although not being able to go as far as to see a process of true substitution of one intensity by the other, it was clear in almost all of the cases that there is an association between the two. The clearest example comes again from Case B and its file of tens of thousands of shade recipes. In addition, Cases C2 and D's business strategies are not feasible without a huge amount of information about consumers trends and needs.

6 – 'Systemation, not just automation, through the uses of information and communication technologies integrated to all major administrative functions within the firms'
Although not as directly connected with the objectives of this research, some of these characteristics were found in the case-studies, for example: the disseminated use of electronic mail within Oilcomp and Chemicomp and the computerized polymerization process in Case D (which certainly could be found in several of the other cases).

From the brief discussion above, the obvious conclusion is that there are several signs of the new paradigm's advance in the chemical and petrochemical industry. The importance of this new paradigm as well as some speculation about its influence upon the industry's future is discussed in the section below.

10.4.2 The 'de-commoditizing' process

The strategic move of the large chemical corporations away from basic commodities towards higher value-added products was briefly reviewed in Section 4.4 and is extensively described in the press and literature in general. It is not clear, however, how far from those bulk products the chemical companies want to be or can be. Certainly they cannot completely eliminate such products from their portfolios. The several acquisitions of fine and speciality producers by large chemical corporations (many of them cited throughout chapters 5, 6 and 7), as well as the 'intensifying R&D and innovation commitment in growing sectors like pesticides and pharmaceuticals' (Achilladelis *et al.*, 1990) may be part of this so-called 'new-paradigm'. What are, however, the implications of such a process upon the chemical commodities?

Bessant (1991) describes some of the effects of what seem to be the new industrial paradigm effects in several process industries such as steel, sugar, paper and chemicals. In all of them he found a pattern of change towards the characteristics described in Section 10.4.1. About chemical commodities he says:

> In bulk chemical business, again traditionally a commodity operation based on scale economy, the same patterns can be detected. For example the supply of carbon disulphide to the textile industry used to be a high volume, low variety business. But, as a recent article points out, '...while a decade ago many companies churned out large quantities of commodity materials generally confident they would find customers for them, increased competition and more fragmented markets have in the past few years forced the industry to steer towards more specialized product sectors in which the focus is on tailoring goods to customers needs'. As the managing director of one company in this business commented, this move has necessitated 'a complete change of philosophy... now we have to be far more marketing-led and offer more of a technical service to customers... now we have a cycle time of a few days and rapid changeover between making different types of material... What we are doing now is really scale-up laboratory work with batches of chemicals going out of the factory in small drums. It's a long way from the days when everything we made at the plant was produced in a continuous stream and went out in tankers.'

Bessant goes still further:

> Similar patterns can be found in the oil industry, where refineries and blending plants are increasingly having to offer smaller batches of customer specific output. For example, the Shell Lubricants plant at Stanlow, UK, offers a full range of lubricants in sizes ranging from bottles of motor car oil to barrels of aviation lubricants. The total range involves some 2500 combinations, made up from over 700 recipes blended and then packed into different packages.

What he seems to be observing is a tacit process of commodity tailoring or customization. Several facts observed during this research seem to reinforce this observation such as:

- the dissemination of strategies of differentiation (even in commodity producers such as Case A);
- the concentration on the higher value-added niches of mature products (such as happened to Case B and its producer-dyed acrylic fibres, and Case C and its branded nylon fibres).

The specialist literature also starts to provide evidences of this product tailoring:

- *Modern Plastics* (1992, pp. 48–50) describes the development of a new sub-group of polyolefins, the 'very low density polyethylene' or VLDPE.
- The influential *European Chemical News* has given several examples of this process. In its issue of 29 September 1991 it describes Neste's investments in polyolefins (typical pseudo-commodities) as the development of a balanced portfolio towards product differentiation. In November 1992 it cites Kenneth Sinclair of SRI International who forecast that a revolution in polyolefins technology (through the use of 'metallocene' catalysts) will rival the gains made in the initial discovery of polyethylene during the 1950s. Among the speciality-polypropylenes to be produced he describes 'ultra-stiff polypropylene' and 'high-clarity syndiotactic polypropylene'.
- An example of commodity-tailoring not related to any polyolefin is titanium oxide which although produced in commodity-scale is sold with speciality's characteristics (*Modern Paint & Coatings*, 1992).

Also in *European Chemicals News*, Eisberg (1992) analyses this whole process of tailoring commodities:

> As the commodity chemicals sector continues to see margins squeezed, producers are beginning to realise that reducing production costs no longer offers a solution. Some now see product differentiation as the only way forward for the industry ...

This paper also cites Duane Dickson, from Gemini Consulting:

> 'Industry wisdom suggests everyone in a commodity concern has pretty much the same price and a commodities product is in itself undifferentiated. The actions of most companies in commodity chemical industry also suggest that being the lowest-cost producer is the best way to dominate. I would argue however that you can drive down costs forever without ever dominating'..., customers are not buying on the basis of cost, since everyone's cost tends to be the same, but rather on the basis of a series of benefits that companies provide. By definition, price becomes neutral.

The paper describes the existence of 'superior value delivering systems', focused on customer benefits at costs that generate acceptable profit.

> This encompasses everything from work processes, the organization and management to control structures, information technology and alignment of the strategy... This means strategy, culture, and leadership will be aimed at delivering value to customers, as measured by the benefits received all the way down the line.

Eisberg condenses the whole idea and identifies it in a single observation: 'The latest buzz-phrase being heard wherever producers gather is de-commoditising commodities.'

While the methodology employed in this book doesn't allow conclusive generalizations for the chemical industry as a whole, it provides important evidence to characterize a 'de-commoditizing' of chemical and petro-chemical commodities. It also provides some important information about what seems to be the first steps of this process, through the 'common-place strategy' of sheer differentiation that it describes. Still reinforcing this idea and providing ammunition for speculation about the future of the industry is the comment of a consultant who was interviewed for this research that in the past the fibres studied here used to be 'commodities' but that now they are 'specialities'. The conclusion is that there are two complementary but different processes taking place in the chemical industry re-structuring: the first is the move away from commodities. The second is a process differentiation through subtle, although valuable, specialization of commodities, the so-called 'de-commoditizing'. Both processes seem to have some of the characteristics of what has been called 'the new industrial paradigm'.

The literature provides a great deal of information about signs of the new paradigm in each of the individual functions which make up a company. Kotler (1992), for example, in analysing 'marketing's new paradigm', defends a stronger link between supplier and customer and describes five levels of commitment, from the 'basic', passing through 'reactive', 'accountability', 'proactive' and ending with 'partnership'.

Although it cannot be expected that very basic commodities such as ethylene or ammonia will one day become differentiated products, there are signs that it may happen to some intermediate products. Furthermore, the managerial practices involved in such differentiation (and exemplified by a move from Kotler's 'basic' classification towards the 'proactive' and 'partnership' stages) can be used even in those basic commodities markets, eventually differentiating producers if not their products.

Among the managerial practices studied in this book are those concerned with the management of technology, business management and their theoretical link which are further discussed below.

10.4.3 The strategic management of technology

From the generic definition of strategy adopted in Chapter 1, the concepts of Business Strategy* and Technology Strategy** were explored in Chapter 2. For all three concepts the definitions adopted were developed consider-

---

* A pattern of resources and competencies deployed through environmental interactions to achieve a business goal or objective.
** A pattern of techno-scientific resources and competencies deployed to achieve a technological goal or objective defined from the analysis of the business's competitive forces.

ing the purposes and methodology to be used in this research. In at least three cases (A, B and C), the managers who took part in the research showed an understanding of and even some intimacy with all the concepts (although not necessarily caring about the exact definition).

The most important conclusion to be drawn from this research is directly connected with those last two definitions; in Case studies B, C2 and D the components of those two strategies could easily be identified but not so easily separated from each other. The resources and competencies deployed to achieve the business goals (in these three case-studies) were technical and scientific in nature, while the technological objectives were also business and marketing goals. In Case B, its focus on dyed fibres, for example, was an obvious choice. Not as obvious, however, is whether its continuous investment (in the technology that allowed that focus) was (or is) a consequence of a business strategy or of a technology strategy since, in reality, it is part of both. The opposite extreme to this ambiguity can be taken from Case C1 where the disharmony between the technology strategy and the effective business strategy resulted in the waste of its research resources and finally in business divestment. A similar ambiguity (in larger or smaller scale) could also be found in Case C2's branded nylon fibres and Case D's synthetic lattices and compounds.

Taking an analysis parallel to that conducted in Chapter 1, which resulted in the description of the concept of 'Strategic Management', this 'very close management of the technology and business strategies' will be referred to as 'strategic management of technology'. It is important to highlight that the use of the word 'strategic' above does not confer any 'quality label' to this practice. As described above it was chosen based upon a parallel conception to that described in Section 1.4; in both cases they resulted from a holistic approach.

From this conclusion about the 'strategic management of technology', it is reasonable to extract a last conclusion from this research. As the three case-studies where the 'strategic management of technology' was observed are also the same case-studies were most of the new paradigm evidences were found, it is logical to suppose that the 'strategic management of technology' and, possibly, the 'de-commoditizing process'[5] are only another 'symptom' of a new industrial paradigm in the chemical industry.

10.5 Summary of conclusions

(1) Technology strategy is usually recognized and fundamentally deliberate (at least in the mature sectors analysed in this research) in the chemical industry.

(2) The coupling of business and technology strategies is widely recognized as very important. All companies studied are highly concerned with this linkage. Different types of approach are in use for promoting this connection. It can be said that all of them are much more concerned with coupling their technology with the strategies of individual busi-

nesses than with broader corporation strategies.

(3) Several managerial techniques and tools are available in the literature to assist the linkage between business and technology strategies. Some were found to be in use in one of the case-studies where they played only a minor role.

(4) Great differences were found between the three different approaches identified in the research. The most similar approaches were found in Cases A and C, both based upon the use of executives whose formal job is to link business and technology strategies. This approach was called 'personnel'.

The second type of approach was identified in Case B and uses an elaborated organizational and 'budgetarial' structure which assures that each individual business supports its own internal technological development. At the same time this approach guarantees a proportional share of expenditure and investment between businesses using the same core technologies in medium-term projects. Lastly this model also ensures that investments in long-term projects are shared among the different divisions or groups of businesses. This model was called 'organizational'.

The last model found is the 'informal' one. This approach was found in a small business in which it is apparently unnecessary to instigate a closer interaction than that already naturally existent in such a small group of managers.

(5) No relation was found between the different approaches identified among the research groups and the types of products produced by the analysed SBUs.

(6) The influence of major corporation strategies upon the coupling of technology and the individual business strategies was perceived simply as 'guidelines'. Nevertheless, the absence of synchronism between the corporation strategies and the individual businesses proved quite dangerous (as observed in Case C).

(7) More sophisticated and formal approaches were found in the largest and most complex organizations.

(8) No evidence of best practice was conclusively observed (the exception was excessive business drive, which stood out as a condition to be avoided). Some hints, however, were gathered and are described below.

The 'organizational' approach seemed to present better characteristics such as a lower dependence upon individual characteristics of executives resulting consequently in more stablity. The existence of centralized research facilities (Case B) seems important for a successful use of this approach as it promotes an exploitation of potential synergies.

The 'personnel' approach although more dependent on individuals and, as a consequence, potentially less 'stable' seems more responsive to changes in strategies.

Lastly, the informal approach seems to incorporate the qualities of the other two. Nevertheless, it is very dependent on the relationship

Approach to the linkage between busines and technology strategies	Where it was identified	Pros	Cons	Observation
Personnel	Huge multi-divisional corporations and large businesses competing in global markets within these corporations.	Tighter control and responsiveness. Seems more appropriate for carrying out deliberate strategies.	May assume too personal characteristics, making its success dependent upon individual performances. May promote instability.	Was observed in two companies whose generic technology strategy approach was classified as 'mixed'. These two companies have decentralised R&D.
Organisational	Medium sized corporation (and business) with a few divisions and common core-technologies.	Has a character more 'institutional' and less dependent upon individuals. Seems to stimulate 'emergent' strategies. Would promote stability.	Not so 'controllable' and responsive to deliberate strategies.	Was observed in a business employing a 'traditional' technology strategy approach and centralised R&D.
Informal	Small business linked to corporations of variable sizes	Very responsive (due to absence of rigid structures). May stimulate a balance between deliberate and emergent strategies.	May prove very volatile and its results will be strongly dependent upon individuals and their relationships.	Was observed in a business apparently changing from traditional to mixed generic technology strategy approach.

Figure 10.1   Pros and cons of the different approaches for promotion of the linkage between technology and business strategies.

between the different departments and individuals, being, conse-quently, volatile and only viable in small businesses and firms.

All these observations and hints are summarised in Figure 10.1.
Some conclusions which were not originally sought are listed below:

(1) There are several signs of the emergence of a new techno-economic paradigm in the chemical industry.
(2) The move of major players in the industry away from bulk commodities has two components: one is the change of their portfolios towards specialities, the other is the 'de-commoditizing' of some of the remaining commodities.
(3) In some cases the business and technology strategies are so interdependent that the differences between them are not evident. This practice, referred to as 'strategic management of technology', seems more evident in those businesses which present more signs of the new paradigm.

10.6 Implications and recommendations

In terms of the existing literature about the new industrial-paradigm (in its broader meaning), it may be thought that the existence of a generic process of 'de-commoditizing' as well as the existence of a 'strategic management of technology' would be investigated in the industry as a whole, as well as a deeper analysis provided of its consequences upon the chemical industry.

Purely speculatively, it can be said that the most important implication of the conclusions described above is that the pattern of innovation, which was described in Chapter 4 as having moved from the hands of individual entrepreneurs to huge corporations, could start moving again. Such a move could be due to a need for 'strategic management of technology' which can apparently (based upon the case-studies here described) be supposed to be better conducted by small and medium-sized businesses and companies. It is important to highlight that such a change would not influence the existing pattern of radical innovations (even though such innovations are not usual in mature sectors) which will more than ever be kept in the hands of large corporations. The resulting picture could be one of successful small and medium-sized companies heavily dependent on incremental innovation resulting from a close interaction between technology and business strategy, and of large corporations still keeping their leadership in terms of radical innovations.

Notes

1. This is not very probable considering the industry's capital-intensiveness and its low attractiveness.
2. Putting this observation together with that about the difficulties of keeping nylon chemicals and fibres as a single SBU reinforces the view that the corporation had decided on divestment a long time before the case-study.
3. It is interesting to note, however, that the conclusion taken in that case-study was that Case A's main advantage over its competitors was related to its size. This advantage, although not related to price, is not linked to the observed characteristics of the new paradigm but rather with the Fordist model.
4. While Case A did not present any important characteristic in relation to this factor, Case C stands out because of its inflexible organizational structure.
5. As it was more obvious in two of these same case-studies (Cases B and C2).

References

Abernathy, W. J., and Clark, K. B. (1985), 'Innovation: Mapping the winds of creative destruction', *Research Policy*, **14**, 3–22.

Abernathy, W. J. and Utterback, J. M. (1978), 'Patterns of Industrial Innovation', *Technology Review* **50** (7), June–July.

— (1979), 'Dynamics of Innovation in Industry', in C. T. Hill and J. M. Utterback (eds.), *Technological Innovation for a Dynamic Economy*, Pergamon, Oxford.

Achilladelis, B. G. (1973), *Process innovation in the chemical industry*, D phil. thesis, University of Sussex.

— (1975), 'History of UOP: From petroleum refining to petrochemicals', *Chemistry and Industry*, 19 April, 337–44.

Achilladelis, B. G, Schwarzkopf, A. and Cines, M. (1987), 'A Study of Innovation in the Pesticide Industry', *Research Policy*, **16**, 175–212.

— (1990), 'The dynamics of technological innovation: The case of the chemical industry', *Research Policy*, **19**, 1–34.

Andrews, K. R., Learned, E., Christensen, C.R., Guth, W. (1965), *Business Policy: Text and Cases*, Richard D. Irving, Inc.

Ansoff, I. H.,(1965), *Corporate Strategy, an Analytic Approach to Business Policy for Growth and Expansion*, McGraw-Hill, New York.

— (1968), 'Toward a Strategic Theory of the Firm', in I. H. Ansoff (ed.) (1969), *Business Strategy*, Penguin, Harmondsworth.

— (1976), *From Strategic Planning to Strategic Management*, John Wiley, London.

— (1979), *Strategic Management*, Macmillan, London.

— (1984), *Implanting Strategic Management*, Prentice-Hall International, New York.

Arni, V. R. S. (1982), *Emerging petrochemicals technology: implications for developing countries*, UNIDO, New York.

Baily, M. N. and Chakrabarti, A. K. (1988), *Innovation and the Productive Crisis*, The Brookings Institution, Washington, DC.

Baker, M. J. (1985), *Marketing Strategy and Management*, Macmillan, London.

Bemelmans, T. (1979), 'Strategic Planning for Research and Development', *Long Range Planning*, **12**, April, 33–44.

Bennet, M. J. and Kline, C. H. (1987), *MIT's Technology Review*, as cited in 'The chemical industry's innovation decline', *Technology Strategies*, December 1987, 19–20.

Bessant, J. (1991), *Managing Advanced Manufacturing Technology*, Blackwell, Oxford.

Bitondo, D. and Frohman, A. (1981), 'Linking Technological and Business Planning', *Research Management*, November, 19–23.

Buck, D. (1989), 'World Textiles: The Future', paper presented at the Textile Institute Annual Meeting, Nottingham, October.

Burchfield, R. W. (ed.) (1972), *Oxford English Dictionary*, Clarendon, Oxford.

Byrne, H. S. (1989), 'Products Research & Chemical: Timely Infusion of Capital Fuels Growth Strategy', *Barron's*, **69**, 58.

Chandler, A. D., Du Pont, A. (1962), *Strategy and Structure, Chapters in the History of the Industrial Enterprise*, MIT Press.

Chemical Economics Handbook (1989), Stanford Research Institute, Palo Alto, California.

Chemical Industry Association (1989), *A Brief Guide*, **2**, March, London.

Chemical Marketing Reporter, (1991), 'Synthetic Rubber Use to Hit 13 Million Tonnes in '95', 18 February, 9.

Chemical Week (1991a), 1 May, 10. (*)

— (1991b), 27 February, 16. (*)

— (1992), 'NGL's tax exempt', 15 January, 26.

Christensen, C. R., Andrews, K. R., Bower, J. L., Hamermesh, R. G. and Porter, M. E. (1965), *Business Policy: Text and Cases*, Richard D. Irving, Inc.

Clarke, K., Ford, D. and Saren, M. (1989), 'Company Technology Strategy', *R&D Management*, **19**, 3, 215–29.

Cook, L. P. and Sharp, M. (1991), 'The Chemical Industry: the Dynamics of Change', in C. Freeman, M. Sharp and W. Walker (eds.), *Technology and the Future of Europe: Global Competition and the Environment in the 1990s*, Pinter, London.

Cookson, C. (1991), 'An experiment in chemical defence', *Financial Times*, 20 June, 7.

Cookson C. and Jackson T. (1991), 'An institution under threat', *Financial Times*, 18 April, 19.

Coombs, R., Saviotti, P. and Walsh, V. (1987), *Economics and Technological Change*, Macmillan Education Ltd, London.

Coombs, R. and Richards, A. (1989), 'The Technological and Business Strategies of Innovating Firms: Sources of Stability and Change', Paper presented to the *British Academy of Management Conference*, September, Manchester.

— (1990), 'The Integration of R&D Strategy and Business Strategy', *CRONTEC, Manchester School of Management*, March, Manchester.

— (1991a), 'Technologies, Products and Firms' Strategies: Part 1– A Framework for Analysis', *Technology Analysis & Strategic Management*, **13**, 1, 77–86.

— (1991b), 'Technologies, Products and Firms' Strategies: Part 2 – Analysis of Three Cases', *Technology Analysis & Strategic Management*, **3**, 2, 157–75.

Cyert, R. M. and March, J. G. (1963), *A Behavioural Theory of the Firm*, Prentice-Hall, New York.

Davies, S. (1987), 'Survey of West European Textile Mill Fibre Consumption by End Use', *Textile Outlook International*, July, 59–69.

— (1991), 'The Man-made Fibre Industry in Western Europe', *EIU Textile Outlook International*, July, 67–79.

Dirks, G. W. (1991), 'Acrylonitrile 1991–1996 in transition', *Dewitt Petrochemical Review*, Houston, 19–21 March.

Dosi, G. (1982), 'Technological Paradigms and Technological Trajectories', *Research Policy*, **11**, 147–62.

Dow (1989), *Annual Report*, Cover and p. 5.

Drucker, P. F. (1955), *The Practice of Management*, W. Heinemann, London.

Durkheim, E. (1933), *The Division of Labour in Society*, Free Press, Glencoe, Ill.

Eisberg, N. (1992), 'Commodities look to specialist future', *European Chemical News*, **19**, 22 June.

Elastomerics (1992), 'Natural Rubber Remains Prime Polymer for Tire Applications', June, 22.

Encyclopedia of Plastics and Polymers (1976), 164–94.

Enos, J. L. (1962a), 'Invention and innovation in the petroleum refining industry', in National Bureau of Economic Research (ed.), *The rate and Direction of Inventive Activity*, Princeton University Press, Princeton, NJ.

— (1962b), *Petroleum Progress and Profits: a History of Process Innovation*, MIT Press.

Europa-Chimie (1991), 8 April, p. d11.

European Chemical News (1991a), 'Commission's tough line on PP price fixing upheld', 4 November, 6.

— (1991b), 'Neste continues three-pronged expansion course', 23 September, 17.

— (1992), 'Catalyst revolution to boost PP demand', 2 November, 30.

Financial Times (1991a), 10 December. (*)

— (1991b), 'Advanced plastics prove too expansive', 23 July.

— (1991c), 21 November, 14.

Fischer, S. and Dornbuseh, R. (1983), *Economics*, McGraw-Hill, New York.

Ford, D. (1988), 'Develop Your Technology Strategy', *Long Range Planning*, 31(5), 85–95.

Freeman, C. (1982), *The Economics of Industrial Innovation*, Frances Pinter, London.

— (1989), 'Technical Innovation in the World Chemical Industry and Changes in Techno-Economic Paradigm', paper for *Lustrum Conference* at MERIT, University of Limburg, Maastricht.

Fryer, C. (1992), 'Losing the way in fibres?', *Chemistry & Industry*, **10**, May, 392.

Fusfeld, A. R. (1978), 'How to Put Technology Into Corporate Planning', *Technology Review*, May, 51–5.

Goldstein, R. F. and Waddams A. L. (1967), *The Petroleum Chemicals Industry*, Spon Ltd, London.

Green, D. (1992), 'Spinning a yarn to its limits', *Financial Times*, 14 January, 12.

Guglielmo, R, (1962), *A petroquimica no mundo*, Ed. Difusao Europeia do Livro, Sao Paulo, Brazil, as cited in M. A. Suarez (1986), *Petroquimica e Tecnoburocracia: Capitulos do desenvolvimento capitalista no Brasil*, Ed. Hucitec, Sao Paulo, Brazil.

Hampel, C. A. and Hawley, G. G. (1976), *Glossary of Chemical Terms*, Van Nostrand Reinhold, New York.

Harris, J. M., Shaw, R. W. and Sommers, W. P. (1983), 'The Strategic Management of Technology', *Planning Review*, January, 29–35.

Henk van den Belt, Henk and Hip, Arie (1987), 'The Nelson–Winter–Dosi Model and Synthetic Dye Chemistry', in Bijker, Hughes, Pinch, *The Social Construction of Technological Systems*, MIT Press.

Hofer, C. W. and Schendel, D. (1978), *Strategy Formulation: Analytical Concepts*, West Publishing Co, St Paul, Minnesota.

Hollander, S. (1965), *The Sources of Increased Efficiency: A Study of Du Pont Rayon Plants*, MIT Press.

Japan Chemical Week (1991a), 'New Solventless Adhesives Markedly Speed Up Lamination For Packages', 14 November, 5.

— (1991b), 'Mitsui Toatsi Adds Synthetic Latex for Coated Paper', 7 February, 2.

Jewkes, J., Sawers, D. and Stillerman, R. (1958), *The Sources of Invention*, Macmillan (rev. edn. 1969).

Johnson, G. and Scholes, K. (1984), *Exploring Corporate Strategy*, Prentice-Hall, London.

Katz, J. (1972), *Importacion de Tecnologia Aprendizaje Local e Industrializacion Dependiente*, Part II, Dep. de Assuntos Cientificos, Secretaria General da O.E.A, Washington, DC.

Kirk-Othmer (1975), *Encyclopaedia of Chemical Technology*, 3rd edition, Interscience, New York.

Kotler, P. (1992), 'Marketing's New Paradigm: What's Really Happening Out There', *Planning Review*, **5**, 20, Sep/Oct, 50–2.

Kuhn, T. (1962), *The Structure of Scientific Revolutions*, Chicago University Press, Chicago.

Lamming, R. and Bessant, J. (1988), *Macmillan Dictionary of Business and Management*, Macmillan, London.

Leadbeater, C. (1991), 'Level of spending on research has far outstripped that of its U.K. Competitors', *Financial Times*, July, 23.

Leaversuch, R. D., 'Add Very Low Density PE to the List of Options in Polyolefin Resins', *Modern Plastics*, **7**, July, 69.

Lewis, J. (1989), 'Trimming the Fat Off the Corporate Body', *Euromoney (UK)*, 127–8 and 132–5.

Liberatore, M. J. and Titus, G. J. (1983), 'Synthesizing R&D planning and business strategy: some preliminary findings', *R&D Management*, **13**, 4 , 206–18.

Linn, R. A. (1983), 'A Sectoral Approach To Strategic Planning For R&D', *Research Management*, January–February, 33–40.

Longley, R. (1991), *Petrochemicals: An Industry and its Future*, The Economist Intelligence Unit, London.

Lorenz, C. (1990a), 'A drama behind closed doors that paved the way for a corporate metamorphosis', *Financial Times*, 23 March.

— (1990b), 'Re-appraising the power base of regional barons', *Financial Times*, 26 March.

— (1990c), 'A cultural revolution that sets out to supplant hierarchy with informality', *Financial Times*, 30 March.

— (1991), 'The trick is to make even the bad times good', *Financial Times*, 4 March.

Macmillan Dictionary of Business and Management (1988), Macmillan, London.

Marquis, D. G. and Mayers, S. (1969), *Successful Industrial Innovation*, National Science Foundation, Washington.

Marx, K. and Engels, F. (1848), *Manifesto of the Communist Party*, London.

Miller, S. A. (1969), *Ethylene and its Industrial Derivatives*, Ernest Benn Ltd, London.

Mintzberg, H. (1978), 'Patterns in Strategy Formation', *Management Science*, **24**, 9.

Mintzberg, H. and Waters, J. A. (1985), 'Of Strategies, Deliberate and Emergent', in A. David and C. Bowman (eds), *Readings in Strategic Management*, Macmillan, London.

Morita, A. (1987), *Made in Japan*, Collins, Glasgow.

Morris, G. (1989), 'Races for Diversification', *Chemical Week*, **144**, 26 , 42–4.

Mulder, K. F. and Vergragt, P. J. (1990), 'Synthetic fibre technology and company strategy', *R&D Management*, **20**, 3, 247–56.

Nelson, R. R. and Winter, S. G. (1977), 'In Search of a Useful Theory of Innovation', *Research Policy*, **5**, 1, 37–76.

Olin, J. (1972), *R&D Management Practices*, Stanford Research Institute, Zurich, 49.

Pareto, V. (1976), *Sociological Writings*, Blackwell, Oxford.

Pavitt, K. (1984), 'Sectoral Patterns of Technical Changes: Toward a Taxonomy and Theory', *Research Policy*, **13**, January, 343–73.

— (1988), Chap.16 in F. J. Van Raan (ed.), *Handbook of Quantitative Studies of Science and Technology*, Elsevier.

Pearson, G. J. (1990), *Strategic Thinking*, Prentice-Hall, London.

Penrose, E. (1959), *Theory of the Growth of the Firm*.

PEP YEARBOOK (1989), Stanford Research Institute, Palo Alto, California.

Perez, C. (1983), 'Structural Changes and the Assimilation of New Technologies in the Economic and Social System', *Futures*, **15**, 4, 357–75, in Freeman (op. cit. 1989).

Peter, L. (1989), 'Today the Friendly Offer, Tomorrow the Hostile Bid', *Euromoney (UK)*, August, 48–57.

Petigrew, A. (1989), 'Longitudinal Field Research: Theory and Method', *BAM Conference*.

Porter, M. E. (1980), *Competitive Strategy*, The Free Press, London.

— (1985), *Competitive Advantage*, The Free Press, London.

— (1987), 'The State of Strategic Thinking', *Economist*, 23 May.

Reuben, B. G. and Burstall, M. L. (1973), *The Chemical Economy*, Longman, London.

Riegel, E. R. (1974), *Riegel Handbook of Industrial Chemistry*, in J. A. Kent (ed.), Van Nostrand Reinhold, New York.

da Rocha, F. (1984), *A Case Study on Learning of Petrochemical Technology: The Propylene Diluent Phase Process*, MSc. Dissertation, Imperial College, London.

Rosenbloom, R. S. and Kantrow, A. M. (1982), 'The Nurturing of Corporate Research', *Harvard Business Review*, January–February.

Rothwell, R. *et al.* (1974), 'SAPPHO Updated – Project SAPPHO phase 2', *Research Policy*, **3**, 3, 258–91.

Rubber and Plastic News (1991), 'Bad Now Better Later', 11 February, 1.

SAPPHO (1972), *Scientific Activity Predictor from Patterns with Heuristics Origins*, Science Policy Research Unit, University of Sussex.

Sethi, N. K., Movsesian, B. and Hickey, K. D. (1985), 'Can Technology be Managed Strategically?', *Long Range Planning*, **18**, 4, 89–99.

Soete, L. L. G. (1979), 'Firm size and inventive activity: the evidence reconsidered', *European Economic Review*, **12**, 319–40. As cited in Freeman (1982, p. 135)

Shota, V. and Bloack, P. M (1987), 'Japan Puts Priority on R&D Creativity', *Chemical Week*, **41**, 5, 31–2.

Shota, U., Brockinton, L. (1988), 'A Maverick Guides Dainippon's Global Strategy', *Chemical Week*, **143**, 6, 22–4.

Schumpeter, J. A. (1947), *Capitalism, Socialism and Democracy*, 2nd edn, Harper & Row, New York.

— (1976), *Capitalism, Socialism and Democracy*, George Allen & Unwin Ltd, London.

Steele, L. W. (1975), *Innovation in Big Business*, Elsevier, New York, 93–4.

Steiner, G. A. (1979), *Strategic Planning*, The Free Press, New York.

Stern, J. P. and Stern, E. S. (1971), *Petrochemicals Today*, Edward Arnold, London.

Stobaugh, R. (1988), *Innovation and Competition: The Global Management of Petrochemical Products*, Harvard Business School Press, Boston, Massachusetts.

Szakonyi, R. (1990), 'Coordinating R&D and Business Planning', *Technology Analysis & Strategic Management*, **2**, 4, 391–411.

Textile Outlook International, 1987, May, 35–5. (*)

The Economist (1982), 'Downstream without a paddle', 9 July, 70–1.

Tregoe, B. B. and Zimmerman, J. W. (1980), *Top Management Strategy*, John Marting, London.

Twiss, B. C. (1986), *Managing Technological Innovation*, Longman, New York.

Walsh, V., Townsend, J., Achilladelis, B. G. and Freeman, C. (1979), *Trends in Invention and Innovation in the Chemical Industry*, Mimeo, Science Policy Research Unit, University of Sussex.

Weber, M. (1962), *Basic Concepts in Sociology*, Peter Owen, London.

Weil, E. D. and Cangemi, R. R. (1983), 'Linking Long-Range Research to Strategic Planning', *Research Management*, May–June, 33–9.

Wey, J., Russell, T. W. F. and Swartzlander, M. W. (1979), *The Structure of the Chemical Industry*, McGraw-Hill, USA.

Wilkinson, A. (1987), 'Corporate strategy as a source of ideas', *R&D Management*, **17**, 1, 51–61.

— (1985), 'Corporate Strategy and the Buying and Selling of Know-how', *R&D Management*, **15**, 4, 261–69.

Wilson, I. (1990), 'The State of Strategic Planning', *Technological Forecasting and Social Change*, **37**, 103–10.

Annual reports of 'Oilcomp', 'Fibrecomp', 'Chemicomp', 'Cheminc', 'Inkco' and 'Britco' from 1972 to 1992. (*)

(*) Titles not cited due to agreement on confidentiality with each company under analysis.